PLANET OCEAN

Translated from *Planète Mers: Voyage au Coeur de la Biodiversité Marine* by Laurent Ballesta and Pierre Descamp, published in France in 2005 by Editions Michel Lafon.

Editions Michel Lafon
7-13 boulevard Paul-Emile Victor - Île de la Jatte
92521 Neuilly-sur-Seine

In collaboration with The World Conservation Union (IUCN)

ISBN: 978-1-4262-0186-8

Library of Congress Cataloging-in-Publication Data

Ballesta, Laurent, 1974-
  [Planète mers. English]
  Planet ocean : voyage to the heart of marine diversity / by Laurent Ballesta and Pierre Descamp. -- 1st U.S. ed.
     p. cm.
  Includes index.
  ISBN 978-1-4262-0186-8
  1. Marine ecology. 2. Marine biodiversity. I. Descamp, Pierre, 1973- II. Title.
QH541.5.S3B3313 2007
577.7--dc22
                        2007008722

Founded in 1888, the National Geographic Society is one of the largest nonprofit scientific and educational organizations in the world. It reaches more than 285 million people worldwide each month through its official journal, NATIONAL GEOGRAPHIC, and its four other magazines; the National Geographic Channel; television documentaries; radio programs; films; books; videos and DVDs; maps; and interactive media. National Geographic has funded more than 8,000 scientific research projects and supports an education program combating geographic illiteracy.

For more information, please call 1-800-NGS LINE (647-5463) or write to the following address:

National Geographic Society
1145 17th Street N.W.
Washington, D.C. 20036-4688 U.S.A.

Visit us online at www.nationalgeographic.com/books

For information about special discounts for bulk purchases, please contact National Geographic Books Special Sales: ngspecsales@ngs.org

Printed in France

PLANET OCEAN

VOYAGE TO THE HEART OF THE MARINE REALM

LAURENT BALLESTA AND PIERRE DESCAMP

NATIONAL GEOGRAPHIC
WASHINGTON, D.C.

WITH THE PARTICIPATION OF THE
THE WORLD CONSERVATION UNION

IUCN
The World Conservation Union

The oceans are vast. But what was once teeming with life, and a seemingly inexhaustible source of food, is now being emptied at a rapid pace. To meet the needs of the growing human population, human beings are taking more and more food from the ocean. The intensification of our exploitation of the sea is depleting its stock of fish and threatening its marine diversity. For many species we have reached, or gone far beyond, the limits of sustainable catches. Newer threats come from climate change and from maritime traffic, which transports invasive species to areas where they may replace economically valuable native species.

From the light-nourished coastal shelves to the deep, dark abysses, from the pelagic open ocean to the giant underwater mountains known as seamounts, the oceans hold a multitude of habitats with a stunning variety of species. The economic value of this ocean biodiversity to the human race is immense and underpins much of our survival. Finding a way to use these resources sustainably represents one of the biggest challenges we will have to face over the next ten years.

The magnificent photographs found in this book should provoke an initial reaction of admiration in the face of such beauty and diversity. Then other questions will surface: How do these organisms live and how do these ecosystems work? What threats do they face?

The World Conservation Union (IUCN) has tapped into the knowledge and experience of its network of marine scientists to bring readers of each chapter an overview of the threats to the marine environment and what is at stake in the search for a sustainable use of "Planet Ocean."

Together with the stunning images of life's beauty in the oceans, I hope this information will also lead you to the question what you can do to protect this valuable heritage to make sure it is still there for future generations to enjoy.

Julia Marton-Lefèvre
Director General of IUCN

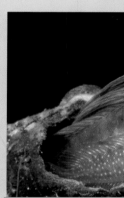

Contents

Preface

This startling, intimate book has the effect of altering time. In fact, for me, this book is all about time. It is about time and, as exotic and inaccessible as its subjects may be to us, it is also a book about us. Let me explain.

Most of us who will be privileged to experience this book lead hectic lives in sizeable cities and speak the fast, telegraphic language of emails. We drive while talking on cell phones in order to multi-task and pack as much as possible into hours that seem shorter and shorter. We are losing the leisurely and calm art of observing the world around us as it unfolds to ancient rhythms. This stunning work of Laurent Ballesta and Pierre Descamp brings us to a screeching halt.

The gift of this book is in the details of a gallery of marine life that even the most experienced diver could never observe. Through state-of-the-art technology, infinite patience, and what must be an intuitive savvy of his subjects, Laurent Ballesta has captured portraits and frozen-in-time behaviors that command the eye to linger—for the pleasure that is art and to ask the questions that arise from such intimate views. Those questions are then expertly answered by Pierre Descamp's clear, brief and well-paced text. It is a brilliant fusion of art and science.

Under water, there is already a different sense of time as all things move through a medium a thousandfold more dense than air. The sensation for a diver is that of altered time as each movement through water passes more slowly. This book's arresting beauty slows us in the same way as the density of water slows us. It gives us time to observe and time to think. My father pushed me overboard on scuba when I was seven and this slower world, where I could watch and think and wonder, became a critical part of who I am. This book puts me in that same reflective, satisfied state of mind. But there is another sense of time captured on these pages and that is evolutionary time.

We suspect that "Planet Ocean" as we barely know it was formed a drop at a time, its vast waters brought as galactic ice on the backs of comets or possibly condensing in sporadic volcanic explosions from deep in the heart of the young planet. As this primordial ocean grew, something unique was sparked, and in this liquid realm life first emerged and immediately set about surviving in as many ways as possible.

This ancient ocean was a laboratory of infinite possibility where life exploded in riotous forms. Sharks were the size of whales, and even the tiniest predators were fanged and clawed. Our closest relatives in this dawn of life were tunicates, short, jellylike tubes glued to rocks and capable only of bobbing in the currents. Unassuming as they were, they were to become the chordates, eventually leading, over evolutionary time, to animals with backbones, including ourselves. But before all that, this sea-beyond-time hosted a vast and colorful carnival, a free-for-all, while the land lay arid and silent and waited.

But we now know that this formative sea runs in our veins, covers 70 percent of our planet, is involved in producing the air we breathe and the water we drink. And, although vast, it is more fragile than we could have guessed.

And here is where we intersect with time and the sea.

This beautiful, smart book is testimony to what we have to gain by living differently on the planet, by changing how we conduct ourselves on land so that these magnificent creatures you are about to meet will continue to thrive and lead lives of their own. Soon, more than half the population of the planet will live within fifty miles of a coastline. Already, ninety percent of large fish have been fished from the sea. We are in the midst of a massive extinction event, both on land and in the sea. We don't know all the answers as to why, but we do know some of the answers about what to do next. And we are obliged to be in a hurry.

Recently, while documenting the value of marine sanctuaries on film, we also documented the value of creating marine reserves, protected oases that dot the marine landscape and where fish can seek refuge and reproduce. Astonishingly, left alone, marine life rebounds quickly. In less than five years, most of these reserves were again flourishing with fish, and fishermen lined the edges of the reserves, catching the overflow in a system that promises to be sustainable, both for the fish and for the fishermen. So we know there are answers to our problems.

On a global scale, we know that we are involved with the contribution of CO_2 in the atmosphere in unprecedented volumes and that the global weather system is reacting, with warming, increasingly intense storms and rapid change along fragile zones such as coral reefs and polar ice. We are being told, in a variety of ways, that the clock is ticking and that time is running out for us to react.

We know that we can live more sustainable lives and that we can learn to change our needs to cause less harm and still live well. This book urgently reminds me why we should.

Planet Ocean is the result, not only of the colossal amount of time Laurent Ballesta spent taking these pictures and Pierre Descamp needed to craft his text, but of the time it has taken for life itself to evolve to the incomparable beauty and significance found in the sea. I hope you will take the time to savor this book and to find meaning for yourself in the short time we all have to make a difference.

Jean-Michel Cousteau

Introduction

"I have a dream
The dream I have
Everyone has
I dream of water
But from the ocean
Ah! The ocean [...]
To live life differently."

Laurent Voulzy and Alain Souchon
The Dream of the Fisherman

"He put on his diving goggles, and for a second a world he had only
imagined in his dreams unfolded before him: a silence rocked by
the waves, a movement transformed into a plant, a light walking
on stilts, spectral colors become fish, when suddenly, a gigantic
pelican, swooped down like damnation, bursting through the roof
of a dazzling sky, and snatched up its prey... but his giggles, which
dated from before the Revolution, filled with water and he had to
return to the sun."

Harry Mulisch
The Discovery of Heaven

At the time of this writing, there is no announcement of a tsunami, marine oil spill, or El Niño, which are so destructive to seas and lives. Would this year finally be free of a major ecological catastrophe? It would almost seem astonishing, so accustomed have we become to suffering a succession of disasters, each of which makes the quality of our environment slightly worse.

To change from these sad, everyday images of oceans soiled by oil spills or seas damaged by overfishing, we have thought of this book as quite the reverse—a blue tide. In our cargo of images, loaded from the four corners of the world's five oceans, we have collected the most beautiful, novel, and fascinating aspects of the marine realm. Eleven years separate the oldest *Planet Ocean* photographs from the newest.

This wealth of images was both simple and difficult to produce. The oceans we think we know hold many treasures to be discovered, and a keen observer is constantly harvesting something new. But the water is also a hostile environment where man is an intruder, and it is rare to be able to stay under water for hours on end. Hidden behind many of the photographs presented here are numb fingers and lips blue from the cold.

In spite of this, and despite decompression stages and seasickness, it would be difficult to measure all the pleasure this adventure has provided us. Pleasure first in observing, then in capturing, fleeting moments. And occasionally, the supreme pleasure of discovering some new secrets of the intimate life of the water's inhabitants. Being able to increase our comprehension of them contributes a tremendous intensity to our observations, which makes it possible to reconcile ourselves to the long afternoons at the university studying, rather than diving. There is no question, however, that even though an underwater picture in and of itself is sometimes enough because of the charm of the animals or the spectacular scenery, often it is nearly incomprehensible. The sight of a kitten in a man's slipper is at first glance more appealing to us than the portrait of a fish taken in the Antarctic.

Thus, the undersea world is so far from our frame of reference that we sometimes become lost in it. Underwater, there are plants that swim, animals that live attached to one place, and numerous mammals that have neither hair nor paws. We must become accustomed to it, or be guided in this separate universe. In a way, the sea requires interpreters. This is why *Planet Ocean* first provides some keys that are essential to understanding marine life, offering introductions to each marine environment. It will then be possible for the reader to embark on a vast underwater voyage of discovery of marvelous undersea landscapes, from icebergs in the polar oceans to coral reefs.

Secondly, *Planet Ocean* is dedicated to the relationships that unite marine animals. From loving to fighting, from collaboration to predation, no species lives in isolation. Finally, the relationships that bind the individuals are often more essential than the living organisms themselves.

Though the forms of undersea life may be totally foreign to us, we still find human attitudes in the behavior of marine animals, with their good points and idiosyncrasies. Of course, no one sees themselves in these animal behaviors; nevertheless, many people will recognize their "neighbors."

In each of the 13 chapters of this book, we wanted to approach the central issues concerning conservation of the oceans, because the inhabitants of the seas—and especially the most charismatic species—are in danger everywhere. The widespread belief that marine species are more resistant to extinction than land species is false. We cannot allow the current deterioration of the marine environment to continue.

The scientific articles in this book have been written by distinguished members of The World Conservation Union (IUCN). They do not pretend to offer an encyclopedic knowledge of the undersea world. Instead, they hope to give readers a general overview of the complexity involved in the conservation of the marine environment, and of the challenge that our modern societies must meet to protect the oceans.

The IUCN texts reflect the diversity and variety of opinions of the people who comprise this vast organization, whether they be partners, members of commissions or employees. The goal of these texts is to present the worries, hopes, and sometimes despair of contemporary leaders in the conservation of the undersea world.

One extraordinary feature of ecosystems is that the properties of the whole are greater than the sum of the properties of the parts. If some of the 400 photos and the 100 anecdotes presented here help put some well-behaved children to sleep without wearying their parents; if some of the 20 texts by IUCN researchers resonate and contribute to changing some mindsets, then our happiness will be complete. Added to the pleasure of producing this book will be the pleasure of sharing it.

The Ocean—
That Great Unknown

The water of the oceans may have originated in outer space. According to scientists, it may have come from comets composed of snow and ice that fell on planet Earth some four billion years ago. Could that be why the marine environment seems so strange to us? It's quite possible, given how going under the surface of the water transports us into another world, governed by other laws. It is not unusual to find plants that swim, or animals that live planted in the seafloor, or fish that turn bright red to disguise themselves.

To become interested in the marine world, one must be prepared to encounter the unheard-of, the strange, the inconceivable. It also means coming face-to-face with the unknown. Covering 75 percent of the planet, the oceans are largely unexplored. From minute bacteria to giant squid nearly 50 feet long, most marine species are still unknown to us. What a place for adventure! So many discoveries, so many new encounters with marine life remain to be made. But to approach this separate universe, which is so close to us and yet still so impenetrable, we need to have some essential guiding information, the first of which is to realize this important fact: The great laws of nature are not the same in the water as on the land.

Living in the Water

Yellowlip emperors (about 14 inches), French Polynesia.

In water, things float. And this changes everything.
Seawater is about one thousand times denser than air.
In this fluid that carries them along, living creatures are,
therefore, not inevitably "nailed" to the ground. The sea angel spends
its life drifting gently with the currents. Moving in such a viscous
environment, on the other hand, is difficult, so most fish,
like the yellowlip emperor, are slender in shape.

Sea angel (about 2.5 inches), Arctic.

Giant Pacific octopus (29.5 ft), Canada.

Do Sea Monsters Exist?

Sunflower sea star (31.5 inches), Canada.

The silence, the shadowy light of the depths,
the proximity of unexplored abysses; the ocean can be conducive to dread.
The repulsive appearance of the giant octopus, the wolf eel,
or giant starfish doesn't help matters.
Despite their looks, they are not at all dangerous for humans.
What if it were just the unknown that we were afraid of?

Wolf-eel
(6.5 ft), Canada.

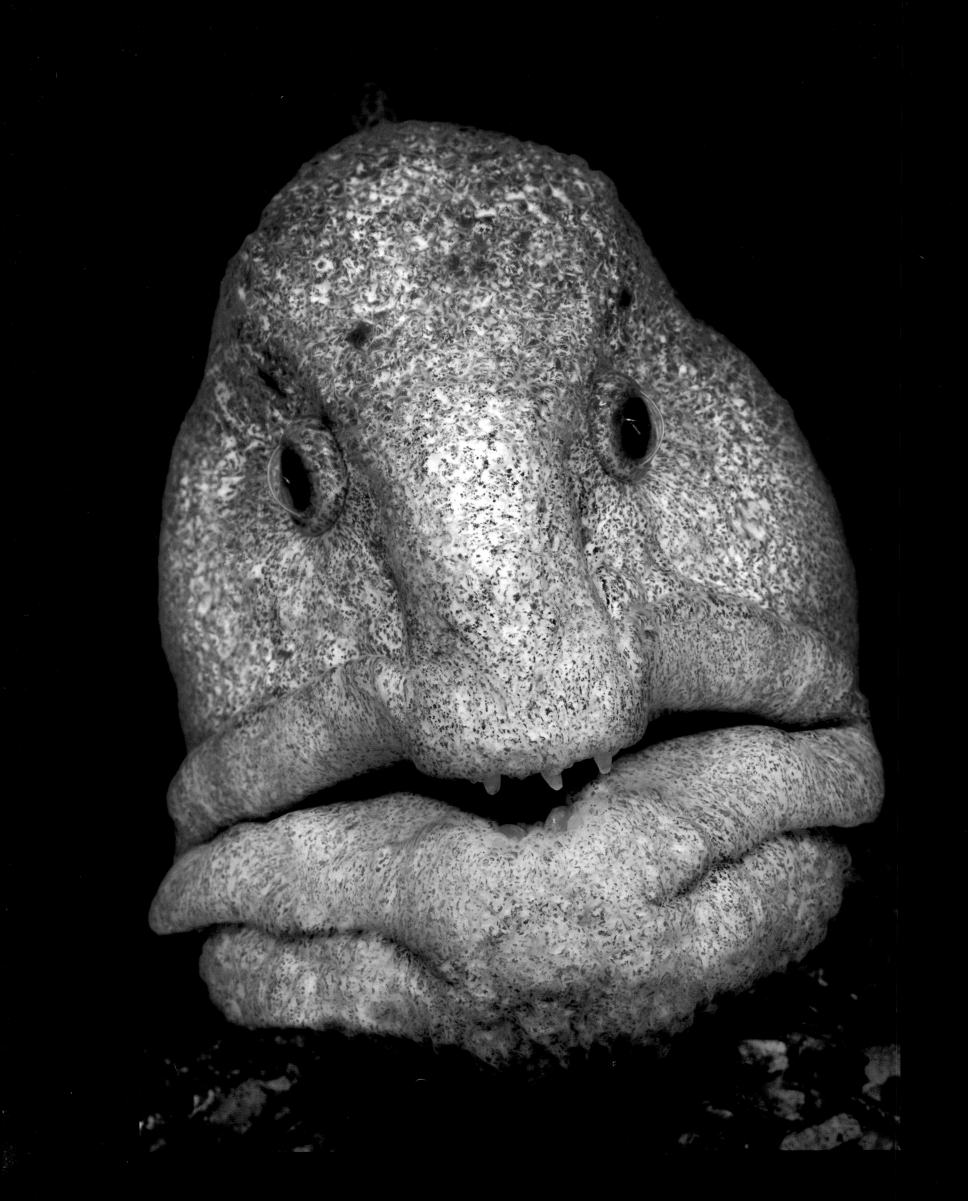

Is the Ocean Dangerous?

Alcyonarian soft coral (27.5 inches), Egypt.

Danger is often not where we expect it.
While hammerhead sharks have never attacked
divers, succumbing to the call of the depths
is far more perilous. To explore the marine world,
divers must display discipline and humility.
Their safety depends more on their proper
knowledge of diving techniques
than on the animals they encounter.

Scalloped hammerheads
(10.8 feet), Mozambique Channel.

Giant plumose anemones (3 ft), and **Sunflower sea star** (31.5 inches), Canada.

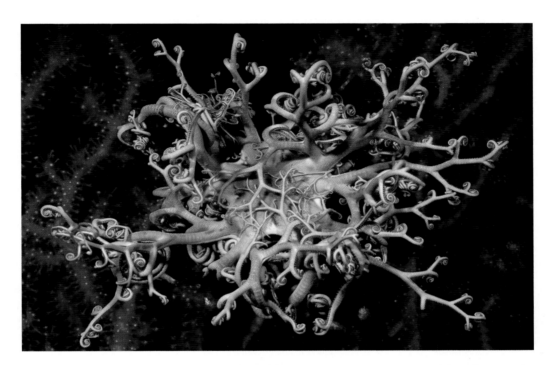

Gorgon's head basket star (13.8 inches), Mediterranean.
Fish-eating anemone (9.8 inches), Canada.

Many marine animals live attached in one place like plants.
The basket star and the anemones need neither eyes to detect their prey nor paws
to pursue them. They require only a trunk or branches to filter the seawater and capture the zooplankton
formed from millions of tiny animals drifting with the currents.

The **northern basket star** (15.7 inches) and **polar alcyonarian soft coral** (12 inches) beneath fissured pack ice, Canadian Arctic.

Why Isn't the Sea Blue?

Sunlight is made up of all the colors of the rainbow.
But as soon as its rays penetrate water, the red, yellow, and
then green are absorbed, and everything appears bluish.
At a depth of about 65 feet, there is no more than about 10 percent
of the surface light left, and a lamp is then needed to make
the true colors reappear.

Red gorgonia (3 feet), **comber** (14 inches) and **swallowtail seaperch** (6 inches), France.

Where Do Waves Come From?

Surf, French Polynesia.

Waves are formed where the wind blows across the surface of the ocean. They can then travel for thousands of miles. In the region of the South Pole, they can even go around the Earth several times. Their height depends on the force of the wind, but above all on the distance over which the wind can blow. This is why waves in enclosed seas, like the Mediterranean, are smaller than waves in the great oceans.

The snake blenny
(23.5 inches), Egypt.

The Andromeda goby
(1.5 inches), France.

The red-necked notothen (3 inches), Antarctic.

Many species are still undiscovered,
including in the known areas of the oceans. The Andromeda goby
is a good illustration of this: It has been observed only once, in 1997—
just a short distance from important marine biology centers that have
been active for more than a century. From the nocturnal serpent blenny
to the red-necked notothen living in the Antarctic and the enigmatic
Oates' soft coral crab, how many species are there
about which nothing is known, except the fact that they exist?

Oates' soft coral crab / **Dendronephthya crab** (0.6 inches), Papua New Guinea.

Red coral
(7.9 inches), Mediterranean.

Sea whips and red coral
are composed of thousands of "persons," polyps
that resemble little anemones and share the same
crimson skeleton. For these creatures, it is very difficult to talk about individuality.
They are colonies of Siamese twins!

Siamese Twins

Sea whips
(4 feet), Papua New Guinea.

The Secret Language of Marine Animals

Dusky grouper (3 feet), Spain.

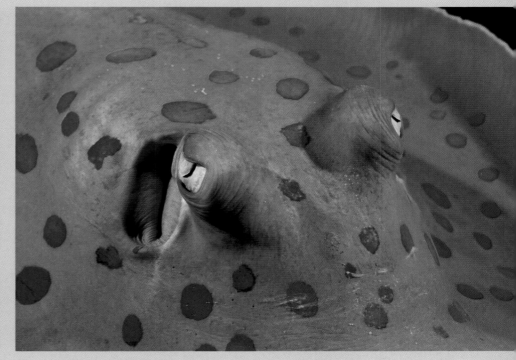

Bluespotted ribbontail ray (3 feet), Egypt.

Do dolphins talk? No one knows.
What is certain, though, is that marine animals exchange signals.
This distinction is important: A word makes one think, a signal makes one react.
The animals' communication can take many forms: chemical, visual, auditory, tactile, or electric.
A grouper's nostrils are extremely sensitive for detecting odors, whether of some prey or of a potential mate.
As for the bright colors of the bluespotted ribbontail ray, they warn any predators
that the ray has a fearsome, venomous stinger.

Bottlenose dolphins
(13 feet), French Polynesia.

The Treasure Hunt

Wreck of the *Schiaffino XXIV,* a 250-foot cargo vessel, France.

The most fabulous of all museums lies under the water. In the roar of storms or battles,
tens of thousands of ships have disappeared over the centuries, carrying their cargo
to the secrecy of the depths. Motivated by an interest in history, a taste for adventure, or a desire for profit,
treasure hunters are legion. And with advances in diving techniques,
the area from 300 feet to 500 feet down is now open to them.

Wreck of the *Bengasi,*
a 280-foot cargo vessel, Italy.

Ocean Biodiversity: Key to Our Survival Strategy

Why should we protect marine biological diversity? Apart from its intrinsic and aesthetic value, there are many vital products that are derived from the oceans. Marine fisheries are an essential staple food for a large part of the world's population and are part of the basis for the culturally associated habits that define us as humans. Marine natural products are another important use of marine biodiversity, and we have only just begun tapping into this field for medicinal purposes or for new sources of food.

The value of marine biodiversity can hardly be overemphasized. Life started in the seas and we are very dependent on the well-being of the seas for our own survival as a species.

The resilience provided by marine ecosystems as a life support for other parts of the world's biodiversity should not be underestimated.

In order to draw the attention of decision makers to the need for improving the way living resources are managed, should we not be using economic evaluation methods to quantify the values of living ecosystems?

Sadly, the reduction of marine biodiversity is likely to continue over the next century. We already know that some extinctions have already occurred, even though our baselines for most groups of species are quite limited. Ongoing habitat destruction and over-fishing is being exacerbated by climate change. In order to address this issue, IUCN and its Species Survival Commission have launched a program to educate the public about the risks of marine extinctions, and what can be done at a practical level to avoid them. IUCN is also involved in a broad range of activities to protect marine biodiversity, from field-based projects and working with local partners, to improving marine resources management to the development of protocols to deal with marine bioinvasions. IUCN strengthens the capacity for science-based management through bringing good science and innovative technologies to the most important management issues. The creation of the IUCN Red List of Threatened species has helped to raise awareness of the dangers facing marine biodiversity.

To end on a positive note, we are seeing the beneficial effects of improved waste water treatment and reduction in pollution from heavy metals in many parts of the world. Management improvements, including the use of Marine Protected Areas (MPAs), are also proving a valuable tool in reducing the damage to fragile marine ecosystems.

The main challenge for the future remains halting the loss of biodiversity in the oceans, so that we can sustain the oceans for the benefit of future generations.

Carl Gustaf Lundin has led the IUCN Global Marine Programme since 2001. He worked as an environmental specialist at the World Bank for more than ten years. A marine biologist, he has traveled the world's seas and oceans to better understand them and to further their protection.

Soft coral cowry (5/8 inch) on an **alcyonian soft coral**, Aqaba Marine Park, Jordan. Today, about 1.5 million species have been identified on land. But our lack of knowledge of the oceans is such that some scientists believe that about 20 million different species that have yet to be inventoried populate the oceans.

Climate Change and Marine Ecosystems

While many effects of global climate change can be observed in the course of our daily lives, some of the most significant impacts are less obvious. Below the surface of the oceans, marine ecosystems are also beginning to show signs of vulnerability to climate change.

From freezing polar waters to balmy tropical seas, elevated levels of carbon dioxide in the earth's atmosphere have initiated lasting changes in environmental conditions that are predicted to have widespread implications for marine life. Among the more significant changes expected are warming sea temperatures, shifts in current patterns, and alterations to the ocean's chemistry.

The food chain under threat

Many marine species are highly sensitive to small increases in average temperature. A prolonged change of only 2°-3° F (1-2°C) can affect growth rates, patterns of reproduction, susceptibilities to disease, and interactions between species. While some species will thrive under warmer conditions, others will decline. The balance between predators and prey, nutrient cycling and other energy flows is likely to be affected, with flow-on effects throughout marine food webs. Where susceptible species play a key role in the ecosystem, such as provision of habitat (e.g., corals) or primary productivity (e.g., plankton), the impacts are expected to flow on to other species that might not themselves be vulnerable to temperature change.

Ocean currents transformed

Ocean currents heavily influence the distribution and productivity of marine ecosystems. Currents transport the young of numerous species, playing a key role in the dispersal and maintenance of populations. Climate change is expected to alter water circulation patterns, leading to changes that interrupt the life cycles of many species and have an impact on local populations, possibility even causing extinctions. Currents also greatly influence nutrient transport. By bringing cooler, nutrient-rich waters to the surface through upwelling, they create areas of heightened biological productivity. Climate change is expected to suppress upwelling in some areas while increasing it in others, leading to major shifts in the location and extent of productivity zones.

Marine chemistry disrupted

Climate change is also expected to alter the chemistry of the ocean. Of particular concern is the decreasing availability of carbonate ions, which are key in the creation of calcium carbonate skeletons for many species. Among these are crucial planktonic organisms, some of which could play important roles in ocean-atmosphere interactions (such as cloud formation). The organisms that build coral reefs also rely on calcium carbonate. As a result, climate change may diminish the processes necessary to maintain the structure that underpins the functioning of coral reef ecosystems.

First signs of evidence

While much of our understanding of the effects of climate change depends on models and scientific predictions, changes are already being observed in some marine ecosystems. Warming of the California Current in recent decades has been linked to declining populations of zooplankton and sooty shearwaters, and shifts toward a predominance of warm-water fish species in kelp forests off the Southern California coast. The spectacular kelp forests off Tasmania (Australia) have declined by 20-80 percent since the 1940s. Dense floating canopies of kelp, they provide a critical habitat for many species. The demise of these kelp forests correlates with an increase of sea temperatures of 2.7°-3°F (1.5-2°C) and a 37-mile (60 km) southward shift in the warm and relatively nutrient-poor East Australia Current.

The marine ecosystems most sensitive to climate change, however, are coral reefs. Corals are particularly sensitive to small temperature changes, and an increase of only 2°-3°(1-2°C) is enough to cause a dramatic stress response known as "bleaching." Bleaching results when the symbiotic relationship between the coral and micro-algae living within its tissues breaks down. Corals can change from their normal healthy color to vivid white within days, and entire communities of corals spanning thousands of miles can be affected for weeks. Entire ecosystems are affected simultaneously, and there are few refuges from thermal stress within shallow reef systems. Even surviving corals suffer sub-lethal effects, and many die if high temperatures persist.

Mass coral mortalities have been reported from around the world in the last decade. Some sites, such as the Seychelles and Maldives in the Indian Ocean, lost 50-90 percent of their corals during the 1998 global bleaching event. Even the vast Great Barrier Reef has been affected, with widespread bleaching (although relatively low mortality levels) recorded in 1998 and 2002. While these temperature-induced events can result in severe impacts on corals, the flow-on effects throughout the ecosystem may be even more important.

Increased sea temperatures predicted under climate change scenarios are also likely to have negative impacts on other species inhabiting reef ecosystems. Unusually high temperatures have caused mass starvation of seabirds on the Great Barrier Reef. Already-threatened sea turtle populations may also be affected by climate change, as the gender ratio of hatchlings is temperature-dependent.

Marine ecosystems, such as coral reefs and kelp forests, are already showing signs of vulnerability to climate change. Yet we continue to struggle to understand the full implications for ecosystems and human societies. By taking the time to look beneath the surface of the oceans, we get a glimpse of the widespread and lasting impacts that await us if the issue of global climate change is not adequately addressed.

__Paul Marshall__ lives in Townsville, Australia. A marine biologist specializing in the ecology of coral reefs, he works for the Great Barrier Reef Marine Park Authority as manager of the Climate Change Response Program. He advises the IUCN Global Marine Programme on issues related to climate change in coral reef environments.

It is estimated that one third of its chemical pollution in the Mediterranean arrives by air, drops into the sea through the rain and then becomes concentrated in the food chain.

Long-snouted seahorse (16 cm) in eelgrass bed, France.

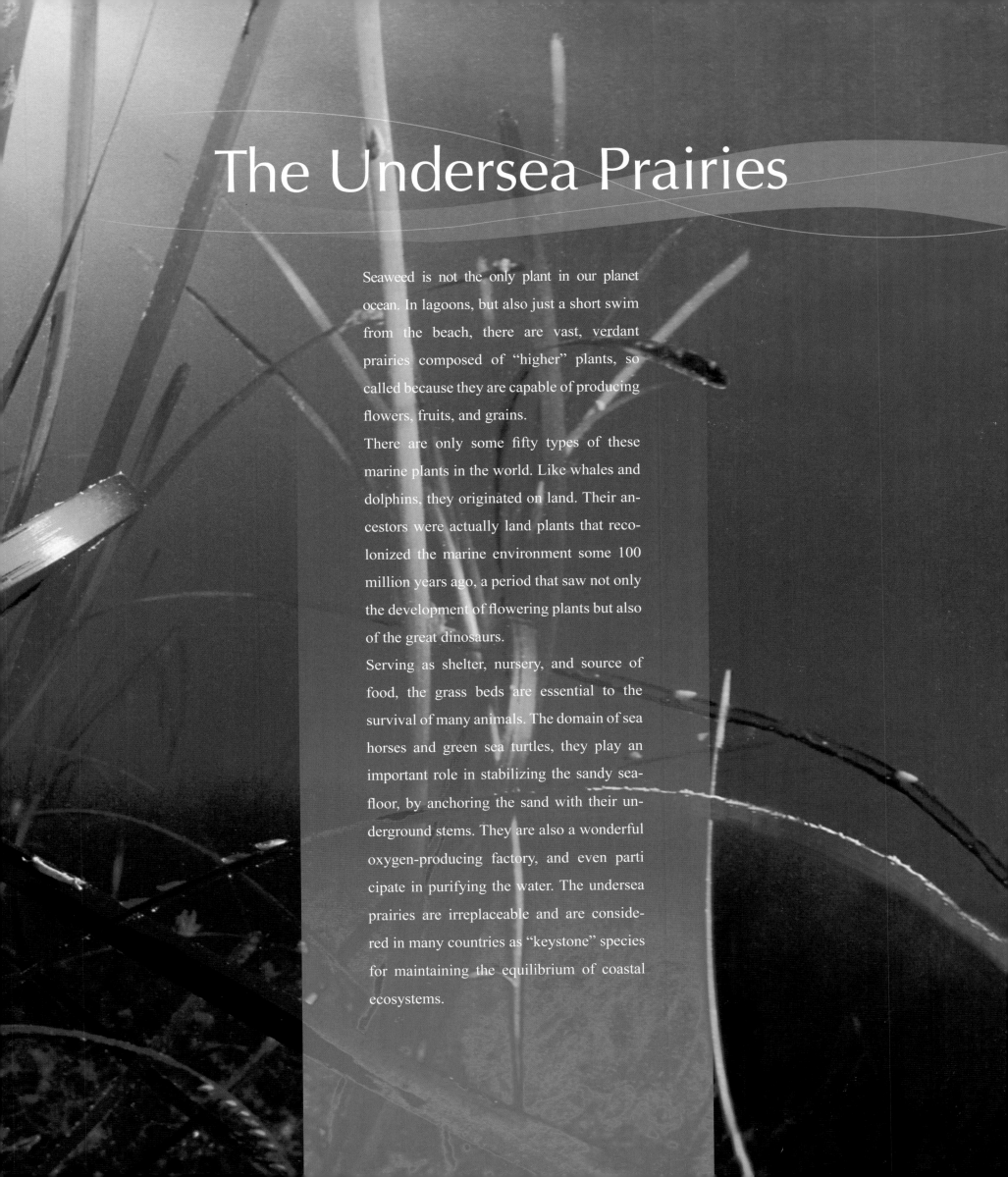

The Undersea Prairies

Seaweed is not the only plant in our planet ocean. In lagoons, but also just a short swim from the beach, there are vast, verdant prairies composed of "higher" plants, so called because they are capable of producing flowers, fruits, and grains.

There are only some fifty types of these marine plants in the world. Like whales and dolphins, they originated on land. Their ancestors were actually land plants that recolonized the marine environment some 100 million years ago, a period that saw not only the development of flowering plants but also of the great dinosaurs.

Serving as shelter, nursery, and source of food, the grass beds are essential to the survival of many animals. The domain of sea horses and green sea turtles, they play an important role in stabilizing the sandy seafloor, by anchoring the sand with their underground stems. They are also a wonderful oxygen-producing factory, and even participate in purifying the water. The undersea prairies are irreplaceable and are considered in many countries as "keystone" species for maintaining the equilibrium of coastal ecosystems.

Vast Pastures

Green sea turtle (6 feet), Mozambique Channel.

Not many herbivores graze in these grass beds.
The leaves here are indigestible!
Only a few marine animals, such as the
Mediterranean salpa or the subtropical green sea
turtles are able to digest them. However, when the
leaves grow old, they are attacked by
hordes of microorganisms that decompose
them and make their organic material
available once again to the food chain. So, directly
or indirectly, the grass bed feeds many
thousands of animal species.

Salpa
(14 inches), France.

Mediterranean tapeweed (1.5 inches), France.

Why do plants produce flowers in shimmering colors?
To attract insects or birds, which, in exchange for a small ration of nectar,
are responsible for disseminating their pollen. So even though herbivores on land eat plants, the plants are
manipulating the animals for their benefit. Marine plants, however, do not use animals to reproduce. That is
why they rarely have flowers and when they do, they tend to be of a subdued green color.

This Is Not Seaweed

Fruit of the southern tapeweed (3/4 inch), Australia.

Red Sea garden eel (27.5 inches), Jordan.

The Sentries of the Grass Beds

Face of the spotted garden eel (27.5 in), Jordan.

When the tidal current rises,
sweeping along its myriad planktonic creatures,
the garden eels come out of their burrows
and draw themselves up to capture their food.
At the least sign of danger, they burrow
back into their deep lairs.
Eels are extremely cautious. It required three days
and two nights before they became accustomed to the
presence of the camera equipment and left
their hiding place to allow this shot,
taken from several yards away by remote control.

1.   Setting up the camera equipment in the grass bed.

2.   Installation of the remote control.

3.   Remote triggering to take the photograph.

1

2

3

Where Does the Pipefish Hide?

The greater pipefish (15.75 inches), France.

Mediterranean tapeweed (3 feet), France.

Pipefish are mimetic.
That is, they resemble their surroundings.
In the picture of the grass bed, can you find
the broad-nosed pipefish that is hiding among
the leaves to surprise its prey?

Long-snouted seahorse (6 inches) in eelgrass bed, France.

Noble pen shell (35 inches), Spain.

The pen shell lives planted in the sand, point downward,
near the grass beds. This 35-inch shellfish, the largest
in the Mediterranean, provides an ideal refuge for the pen shell shrimp
that lives inside the mollusk, well protected from predators.
This unassuming guest does not bother its host much.
The shrimp is satisfied just to divert for itself a little of the plankton
filtered by the pen shell.

Skeleton shrimp
(3/4 inch), France.

Clinging to the Branches

For many of the grass bed's inhabitants, clinging securely to the leaves is vital in order to withstand the force of the swells. Suckers, claws, pincers—anything will do to hold on. But the prize for adaptation goes to the seahorse, which has a prehensile tail that it uses to wrap around the leaves.

Long-snouted seahorses (6.3 in), France.

Sea slug (1.2 in), France.

Sustainable Fisheries: A Long Way to Go

The sea is a vitally important resource, which contributes to the food, economy and health of many nations. Stocks of fish and shellfish can renew themselves if they are allowed to mature, spawn, and produce young in sufficient numbers. Sensible management, with the cooperation of fishermen themselves, can reduce the impact of fishing and ensure that exploitation of marine resources remains sustainable into the future.

We still have a long way to go. Fishermen have become more efficient as they buy larger boats and invest in new methods for locating and catching fish. Modern technology allows for fish to be caught faster than they are able to reproduce. As fish cannot be claimed as property by any individual fisherman or community until they are caught, the number of fishermen tends to increase, and a race for fish develops.

If it is left unregulated, fishing usually results in overinvestment in ships and processing facilities. Costs rise, while incomes, which are determined by shrinking stocks of fish, diminish. Governments are then tempted to provide subsidies, which encourage even heavier exploitation of the fish stocks. Unless strict controls are imposed on fishing efforts, the end result is a decline in both the fish stocks and the fishing industry.

Fishing also has an effect upon the wider ecosystem. Some fishing gear, such as that used for trawling and dredging, may degrade habitats and destroy vulnerable flora and fauna. Furthermore, the removal of fish and other organisms, both deliberately and accidentally (as by-catch), affects the diversity of the ecosystem and the abundance of species such as seals, cetaceans, and seabirds.

Getting fishing under control

Fisheries are managed by national governments in areas of national jurisdiction (which may extend out to 200 nautical miles). Beyond that limit, responsibility switches to regional fisheries management organizations (RFMOs) and flag states. Regulations can be put in place to reduce the capture of young fish and to limit the capture of spawning adult fish. These may include fishing licenses, quotas on the quantities of fish that may be landed, limitations to the size of vessels and the types of fishing gear that may be used, closed areas or seasons to protect juvenile or spawning fish, and restrictions on the number of days fishermen can spend at sea. These measures often demand a heavy investment by governments in collecting scientific information on the state of fish stocks, agreeing on the control measures, and then policing the regulations introduced.

Fishermen often do not accept the rules and regulations imposed as they see their livelihoods affected and often do not trust the scientific opinion. At the same time, environmental groups are asking for ever-stricter measures to protect marine resources, habitats, and ecosystems. There is no trust and practically no dialogue among the various stakeholders and serious conflicts occur. Consequently, reaching some agreement on the proposed measures becomes difficult if not impossible at a national level. At the international level, where multiple nations compete for the same resource, requiring international cooperation, the task is even greater.

There are no simple answers to the problems of managing fisheries. The issues are essentially political, and are about the management of people rather than the management of the biological resources. There are, however, some key factors, which, if applied, can contribute to ensuring sustainable fisheries:

- Reliable scientific data
- Protection of the overexploited stocks with appropriate recovery plans for depleted resources
- Reduction of by-catch
- Protection of valuable habitat
- Effective control, good enforcement, and the elimination of illegal fishing, which at present is the biggest reason for the depletion of some stocks
- International cooperation in managing fish stocks in international waters as well as fish stocks that migrate or straddle between national jurisdiction and the high seas.

Towards good governance

Good governance is a critical goal. All stakeholders need to be consulted, particularly those most affected. To this effect, fishermen need to be involved in all levels of decision-making, because they are the ones that apply the measures at a practical level. Only on finding a consensus between all the interest groups, managers, fishermen, and conservationists will we ensure that fisheries resources are exploited in a sustainable way and vulnerable habitats and species are protected. There is still a long way to go before we can be sure that future generations will have access to the same resources that we harvest today.

Despina Symons Pirovolidou is head of the European Bureau for Conservation and Development in Brussels (a nongovernmental member of IUCN). She chairs the Working Group on Fisheries for the Sustainable Use Specialist Group of the IUCN Species Survival Commission.

The overfishing of bluefin tuna in the western Atlantic in the 1980s led to an estimated 80 percent drop in its tuna stocks. Twenty years later, despite the establishment of quotas, the stocks in this area still have not recovered.

What Will Tomorrow's Aquaculture Look Like?

The term "sustainable aquaculture" encompasses three principal criteria: environmental concerns, societal issues, and economic development.

Rapid growth

Declining natural stocks and a reduction in wild fisheries activities have provided the natural stimulus for the development of aquaculture—particularly what the industry refers to as "finfish culture." Although freshwater carp culture has been practiced for centuries, trout farming only really took off in the 1960s. In Europe, for instance, the introduction of Atlantic salmon culture in Norway and Scotland led to a rapid development of the sector, mirrored later by the rearing of European sea bass and gilthead sea bream in the Mediterranean area. Turbot is also cultivated successfully, and the latest activities include farming Atlantic cod, in northern Europe, and fattening bluefin tuna, in the south of Europe. In only 20 years, marine finfish culture increased from 20,000 tons (1982) to 890,000 tons, according to the Food and Agriculture Organization of the United Nations.

The speed of this growth has inevitably raised specific concerns and issues for this sector.

- **Competition for space**. There is keen competition for the use of coastal areas—for private and public purposes. Debate has started on the possibility of offshore aquaculture, some distance from the coast. At present, there is inadequate knowledge on the types of structures that could be used or on how such sites could be managed practically.

- **The environmental impact**. While all aquaculture units undergo an Environmental Impact Assessment for license approval, considerable research and management efforts are made to reduce environmental impact, particularly regarding the release of waste materials.

- **Healthy reproductive stock and offspring**. Maintaining a healthy stock of cultured fish is paramount for sustainable aquaculture. The lack of licensed therapeutic agents to act against diseases and infections, while assuring the reproduction of high-quality offspring, is a significant problem. While part of this problem is legislative, there is a lot of promising research work being done on improving the inherent quality of offspring, without use of GMO technology (genetic modification).

- **High-quality feeds**. Many of the different cultured species are carnivorous—and therefore need high-quality protein and fats in their diet. Their feed has traditionally been obtained from unused by-catch from fishing activities (so-called trash fish). If global aquaculture growth is to be assured, alternative sources of this feed need to be developed and applied.

- **Pressure on prices**. Increases in the costs of production have not always been reflected in an equivalent rise in the market price for farmed fish.

- **Traceability**. It is in the interests of aquaculture producers to present to society a transparent professional sector that assures quality and full traceability. Many efforts are being made to differentiate regional fish culture products, whether by clear labeling or the adoption of organic farming principles.

To achieve further growth in the aquaculture sector, investment is required, and to obtain that investment, profitability needs to be assured. It has been demonstrated that each job created in European fish culture creates or supports 25 other jobs, thus its contribution to employment cannot be ignored, particularly in rural coastal areas.

Finding the right balance between technologies and environmental management systems, the distribution of products at fair and equitable prices, the maintenance of high-quality products, the promotion of best practices, and the management of the sector's contribution to society, all need to be achieved in a spirit of transparency and good governance. Successful achievement of these tasks will ensure the sustainable development of finfish culture at an international scale.

Courtney Hough is Secretary General of the European Federation of Aquaculture Producers (FEAP), the first such organization to develop a code of conduct for the profession. Prior to that, he worked on aquaculture development in Africa, Latin America, Europe, and the Caribbean. He is also collaborating with the IUCN through the Memorandum of Cooperation between FEAP and IUCN.

Salmon breeding, as it is currently practiced in farms in outdoor settings, constitutes a serious threat and is causing a decrease in wild salmon populations, through hybridization, the introduction of new parasites, and disturbed migratory routes.

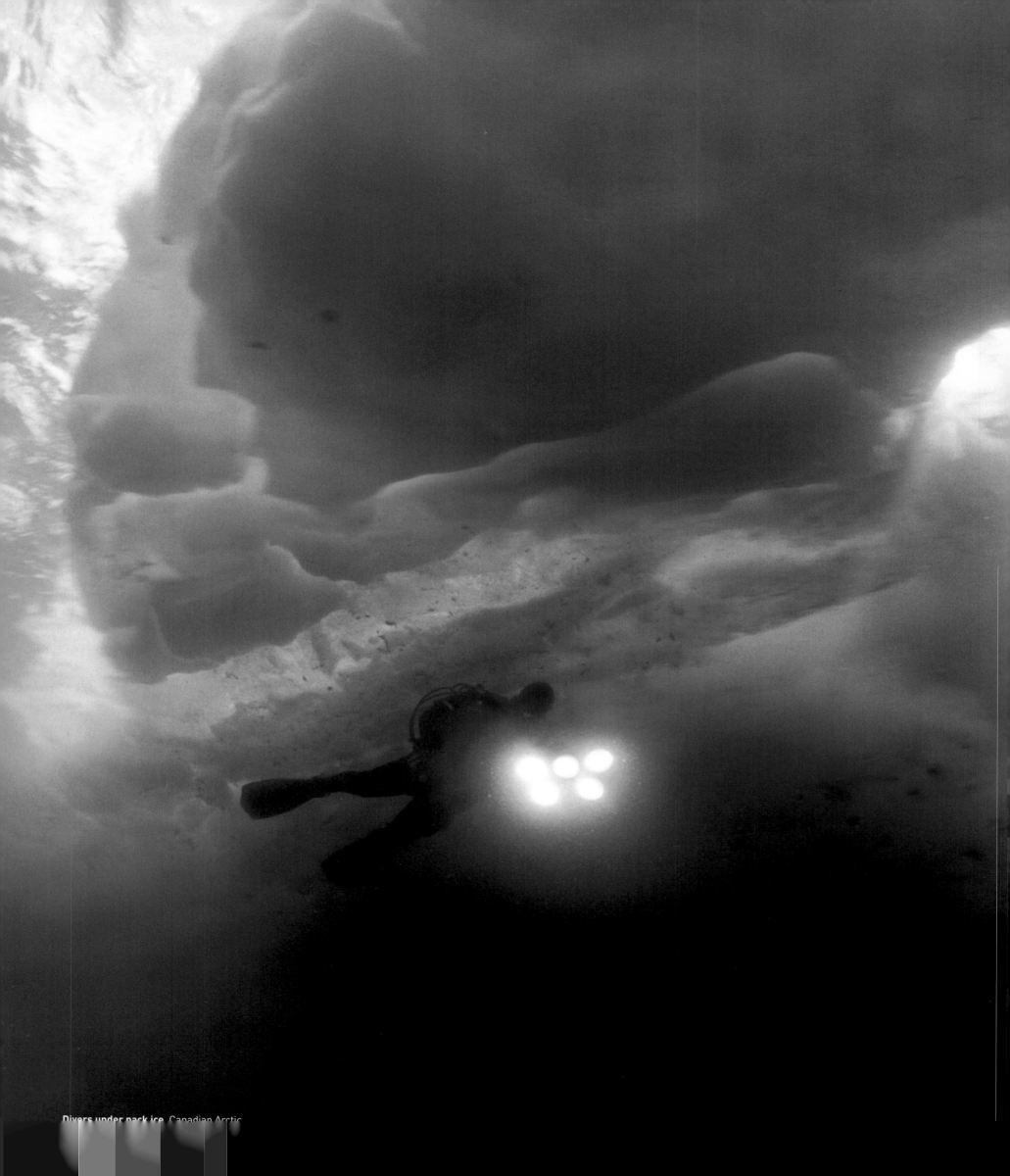

Divers under pack ice, Canadian Arctic

The Polar Oceans

The polar oceans receive as many hours of sunlight as the tropical regions. It is not that the length of time that they receive sunlight is shorter, but rather that they receive less energy from it. The sun shines 24 hours a day during the summer but stays low on the horizon, then sets in the autumn for a night that lasts for six months.

At the poles, life reaches its limits. Outside the water it is practically nonexistent. No organism can withstand the -112°F of the South Pole during the winter. Living beneath the water is more comfortable: the temperature is stable, although always low; there are no blizzards; and there is not much competition, since so few species have succeeded in adapting to the cold.

Life at the poles is therefore essentially marine life, and its richness is as surprising as it is difficult to access. Invertebrates, fish, and polar animals are characterized by a slowness that is not without a certain serenity. Here the environmental stresses are so powerful that they struggle more against the cold than against competitors. Their lethargy helps to preserve their strength. Welcome to the realm of slow action and passive resistance.

Antarctic sea stars (8 inches), Antarctica.

Six Months of Life, Six Months of Night

Sun sea star (15.75 inches), Antarctica.

During the six-month polar night, seaweed stops carrying out photosynthesis. An essential link in the food chain breaks, and the amount of food available to animals becomes very limited. To survive such a night, sea stars and sea urchins concentrate their life cycle over the summer period. During this short time, they reproduce and rush in great numbers on the slightest speck of food in order to accumulate enough reserves for winter.

Greenland shark (16.5 feet), Canadian Arctic.

The Country of Placid Animals

Greenland shark (16.4 feet), Canadian Arctic.

In the glacial waters of the poles,
lethargy is appropriate. Fish, sea slugs,
and whelks all exhibit the same slowness.
The cold slows down the metabolism,
and each of these creatures is conserving its energy.
The growth of polar animals is extremely slow. A
Greenland shark captured 16 years after having been
tagged had only grown three inches.

An **arctic whelk** (2 inches), Canadian

A large **Arctic sea slug**
(6 inches), Canadian Arctic.

Narwhals (13 feet), Canadian Arctic

Computer-synthesized picture.
Virtual representation of an iceberg.
(height: 165 feet, depth: 1,475 feet)

An iceberg's underwater ice cliffs,
Antarctica.

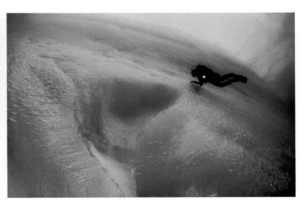

There are two types of ice: icebergs, which are composed of freshwater and produced by the flow of polar glaciers into the sea, and pack ice, which results from the freezing of the ocean's surface. Nine-tenths of an iceberg is submerged. Beneath the surface, the water sculpts the ice, which takes on the appearance of a giant golf ball. In the labyrinth formed by the breakup of the arctic pack ice, narwhals go back to their summer feeding areas.

The Hidden Face of an Iceberg

Antifreeze in the Blood

Antarctic dragonfish (20 inches), Antarctica.

The temperature of the polar seas hardly varies at all.
Throughout the year it stays between 32°F and 28.8°F.
To survive such cold, the blood of the Antarctic dragonfish
is loaded with glycopeptides, "antifreeze" molecules that prevent
ice crystals from forming in its blood.

A **notothen / rockcod** (15.75 inches), Antarctica.

Extreme Environment, Extreme Adaptation

The naked dragonfish (16.5 inches), Antarctica.

From having antifreeze in their blood to hibernating, the fish at the poles dedicate
much of their energy to their struggle against the cold. Nevertheless,
here as elsewhere, carnivores have to hunt while at the same time escape from
their predators. The naked dragonfish adds some spectacular weapons
to its exceptional resistance abilities: two spines located behind its head
make it uneatable, and its protuberant teeth allow it to grab its prey.

Giant Antarctic isopod (6.7 inches),
Antarctica.

Rayed anemone (10 inches), Antarctica.

When a species is particularly well adapted to its environment,
the average size of its members tends to increase over the generations.
This is the reason groups such as isopods, some families of fish,

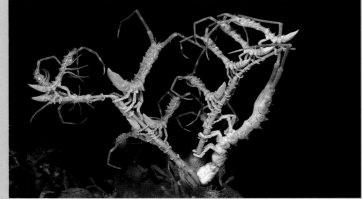

Arctic isopod (6 inches), Canadian Arctic.

This acrobatic isopod has family spirit.
In order for its little ones to be able to feed by filtering the seawater,
it carries them on its antennae, an ideal means to
expose them to the nutritive flow of plankton swept along by the currents.
By doing this, the isopod also protects them
from the many predators that could attack them in the first
fragile stages of their life.

The Acrobat Family

Polar Oceans and Their Unknown Ecosystems

Arctic Ecosystems

Arctic marine biodiversity is unique, consisting of special habitats—from deep ocean basins (typically 2–2.5 miles in depth) to wide and shallow shelf regions—and species that have evolved to live under the harsh Arctic environmental conditions.

Life in the Arctic requires special adaptations to cope with the extreme weather. Birds and mammals are warm-blooded animals that are faced with the challenge of reducing heat loss to keep their bodies warm. All other organisms in the Arctic are cold-blooded; for them the main challenge is to prevent their body fluids from freezing when temperatures dip below the freezing point for seawater (about 28ºF).

As for plants—principally tiny phytoplankton cells or algae growing on the under-surface of the ice—their level

In less than thirty years, the Arctic sea-ice has lost about 40 percent of its thickness and 8 percent of its surface area, reducing the hunting grounds available for polar bears, already greatly threatened by chemical pollution.

of primary production is generally low and limited to a short summer period.

As the ice retreats northward from its maximum distribution in late winter, a sweeping band of phytoplankton production follows that, in turn, triggers and feeds the reproduction and growth of Arctic zooplankton, of which copepods are the dominant herbivores. Relatively large, 5/32–5/16 inch in length, they "fuel up their tanks" with a high number of lipids, an energy store for the winter. The high lipid content of zooplankton is conveyed to their consumers, contributing to the overall high importance

of lipids in Arctic marine food webs. Seals and whales use lipids to form an insulating layer of blubber that reduces heat loss and acts as an energy store for the long winter period without food.

Seasonal migration

About half the surface area of the Arctic Ocean (about 8 million square miles) is covered by seasonal ice in winter and turns into open water when the ice melts in summer. This dynamic sets in motion large migrations both within and into and out of the Arctic region. Each spring and summer very high numbers of animals arrive to feed and grow during the short and hectic Arctic summer. This includes shorebirds and waterfowl, such as geese and ducks, which use the Arctic as their breeding area. It is also the case for the large whales that come to feed on zooplankton and small fish before returning to warmer waters at lower latitudes, where they spend the winter and reproduce.

The true Arctic residents among mammals and birds also migrate in response to the seasonal retreat and advance of the sea-ice. Three whale species are residents of the High Arctic: the bowhead whale, the white whale or beluga, and the narwhal. Large seabird populations also occur in the Arctic, feeding on small fish and zooplankton.

Seals are important components of the Arctic marine ecosystem. The ringed seal, in particular, is a key species in these waters, forming the main prey of polar bears. Ringed seals in turn feed on polar cod and amphipods, which are other important components of High Arctic marine food webs. The walrus, also a common species of the High Arctic, feeds mainly on shellfish and other bottom-dwelling animals.

Under threat

The unique marine biodiversity in the Arctic has not been properly catalogued and mapped, yet it is now under threat from a number of pressures.

Climate change

During the course of this century, much of the Arctic Ocean may end up being open water during the summer. This may have dramatic effects on Arctic marine life. Species such as polar bears and ringed seals may be threatened with extinction since they depend very much on the presence of ice for their subsistence. Other populations, by contrast, may find conditions improving and may extend their range northward under a warmer climate.

Pollution

Persistent Organic Pollutants (POPs) are a serious threat.

These substances (pesticides, chemicals, etc.) are transported northwards via winds, rivers and ocean currents. Here they are taken up in organisms, accumulate, and become concentrated in the species at the top of the food chain. This is the case for PCBs and DDT in the lipid-rich tissues of polar bears and the glaucous gulls.

Oil and gas exploitation

With the increase in demand for energy, there is likely to be growing pressure to exploit the Arctic's important reserves of fossil fuels. There will be an element of risk associated with pioneering developments under harsh Arctic conditions. A major oil spill from a grounded tanker or a blow-out could have grave ecological consequences.

Furthermore, decreased ice cover due to global warming will open up new shipping routes and, with it, more of the dangers associated with maritime transport. Seabirds and mammals would be the first victims.

Overexploitation of living resources

The subarctic contains important fish stocks that support some of the world's largest fisheries. Effects on the targeted stocks, indirect effects on other parts of the ecosystem through food-web interactions, and direct effects on bottom habitats are among the issues of concern.

To address the threats faced by Arctic marine ecosystems, there should firstly be a concerted effort to produce a better inventory and map of Arctic habitats and species. Secondly, an ecosystem-based approach should be implemented for the management of the large marine ecosystems of the Arctic. Finally, the issues of climate change, pollution by POPs, energy supply from fossil fuels, and shipping are concerns that have worldwide consequences and should be dealt with in the proper global contexts.

Hein Rune Skjoldal is a marine biologist. He works for the Institute of Marine Research, in Bergen, Norway, and has led many expeditions in the Barents Sea, the Norwegian Sea and the North Sea. He is currently leading a study on the evaluation of oil and gas exploitation in the Arctic.

Antarctic Marine Ecosystems

Covering a surface area of some 12 million square miles, the Antarctic ecosystem generates some of the richest nutrient conditions on Earth. They sustain an extraordinary concentration of marine species, including charismatic predators at the top of the food chain, such as penguins, albatrosses, seals, and whales.

Food chains

Within the Southern Ocean, the main food web is centered on the shrimp-like crustacean Antarctic krill. This feeds mainly on phytoplankton and is the main prey of many species of seabirds (especially penguins), seals, and large whale species. It also supports exceptionally rich stocks of fish, including the Antarctic toothfish (a kind of cod), the target of a small, but valuable commercial fishery. Krill itself is also the target of a major commercial fishery.

A parallel, but less well-understood Antarctic food web relies on trophic links between phytoplankton, copepod/amphipod communities and their predators (mid-water fish, especially myctophids and cephalopods), which in turn form important prey for a range of species of toothed whales, elephant seals, emperor and king penguins, and certain species of large predatory fish. This web contributes directly and indirectly to the sustenance of the region's most commercially valuable harvested resource, the Patagonian toothfish.

Fisheries management and the precautionary principle

The management and conservation of the Antarctic ecosystem fall under the responsibility of the Convention for the Conservation of Antarctic Marine Living Resources (CCAMLR). Established in 1982 in the wake of serial over-exploitation of fur seals, whales, and Antarctic cod, this intergovernmental organization (with 24 state members) was the forerunner of the use of the precautionary principle in the management of marine systems. It has developed management principles that have resulted in some notable successes:

- a model-based system with explicit precautionary decision rules limiting krill harvesting;

- effective prohibition of fishing on depleted fish stocks to allow their recovery;

- stringent rules for the development of new fisheries that are linked to stock assessments;

- regulations requiring the use of measures to avoid by-catch of seabirds and minimize by-catch of non-target fish species.

However, even within these areas of achievement, CCAMLR has failed to achieve the requisite timing and harvest levels for the krill fishery (due to opposition from the main fishing interests), to restore depleted populations of fish overexploited prior to the organization's creation, and to prevent the development and spread of illegal fishing of toothfish.

Today, CCAMLR faces a number of challenges to adequately protect the marine ecosystem of a region that is of truly global significance, particularly with regard to the management of fisheries, habitats, and the environment. The Antarctic marine system has suffered less environmental degradation than any other marine system on Earth, while oil and mineral exploration is banned for another 30 years. However, the challenge of managing exploitation of marine resources is preeminent. One of the priority requirements is achieving appropriate management of the krill and toothfish fisheries, through the strengthening of control and enforcement measures, as well as better integration of scientific and economic data into management and development models. More generally, environmental issues relating to bottom trawling, especially in relation to seamounts and other sensitive benthic habitats, need to be considered; better standards of environmental performance of fishing vessels need to be developed; and appropriate systems for managing environmentally sensitive areas have to be introduced, including strict protection, where appropriate. Finally, there remain conflicts of interest to resolve between commercial and scientific objectives, especially within the Antarctic Treaty area.

__John Croxall__ is a specialist in marine vertebrates and serves as head of conservation biology at the British Antarctic Survey. He has worked on the management of Antarctic marine resources since 1985 and is a member of the IUCN-WCPA High Seas Task Force.

As a result of global warming, the "permanent" ice of the Antarctic—ice known by that name because it is more than ten thousand years old—is breaking up into immense, tabular icebergs.

Masked crab (2 inches), France.

The Undersea Plains

At first glance, the seabed of sand and silt seems quite peaceful. There is no noise, only a light hubbub of clicking sounds; there is very little movement, or at most only a gentle rocking current. A sensation of calm. Nothing seems to move. And yet everything here is in movement. Nothing is fixed, nothing lasts. Each storm changes the physiognomy of the seafloor by moving millions of tons of sand. For the marine animals, this instability is hellish. Impossible for them to settle themselves in, impossible for them to hide from predators, this is a world without shelter.

In this universe that is constantly being remade, the flora and fauna are not flamboyant. But it would be a mistake to think of these plains as deserts. For although their inhabitants may not distinguish themselves by their bright colors, they are masters in the art of discretion.

At the expense of making incredible adaptations of body forms and behavior, a world of fauna lives here, buried, camouflaged, crouched down. In the country of discreet animals, the beauty of nature goes beyond appearances. Do you know how to discover it?

Wide-eyed flounder
(10 inches), France.

Greater weaver (12 inches), France.

In the plains, having good camouflage is essential.
Predators patrol tirelessly a few inches from
the ocean floor, searching for their food.
This camouflage can take many forms. Burrowing
down, flattening out, taking on the color
of sand—anything to disappear. But in the art of
camouflage, the flatfish reign. Imagine a
deformed-looking fish: a fish lying on its side, its eyes
moved around so that they're both on the same
side of its head, and its body only colored on one side.
Unbelievable, right? And yet that's how flatfish,
such as the flounder, have adapted over the
generations to the motion of the seabed.

Degen's leatherjacket (12 inches), Australia.

A Skeleton Made of Water

Red cedar forest, Canada.

Even the slightest storm can cause the
sandbanks to move and the seabed to give way,
so staying in place is quite an achievement.
The sea pen has a considerable asset
to help it meet this challenge: a skeleton made of
water. It buries its long foot in the sand,
then fills up with water. Under the pressure,
its foot gets a stronger anchorage while its body
stands up, ready to filter the seawater
and capture plankton.

Orange sea pen
(18 inches), Canada.

Stingray (24 inches), Polynesia.

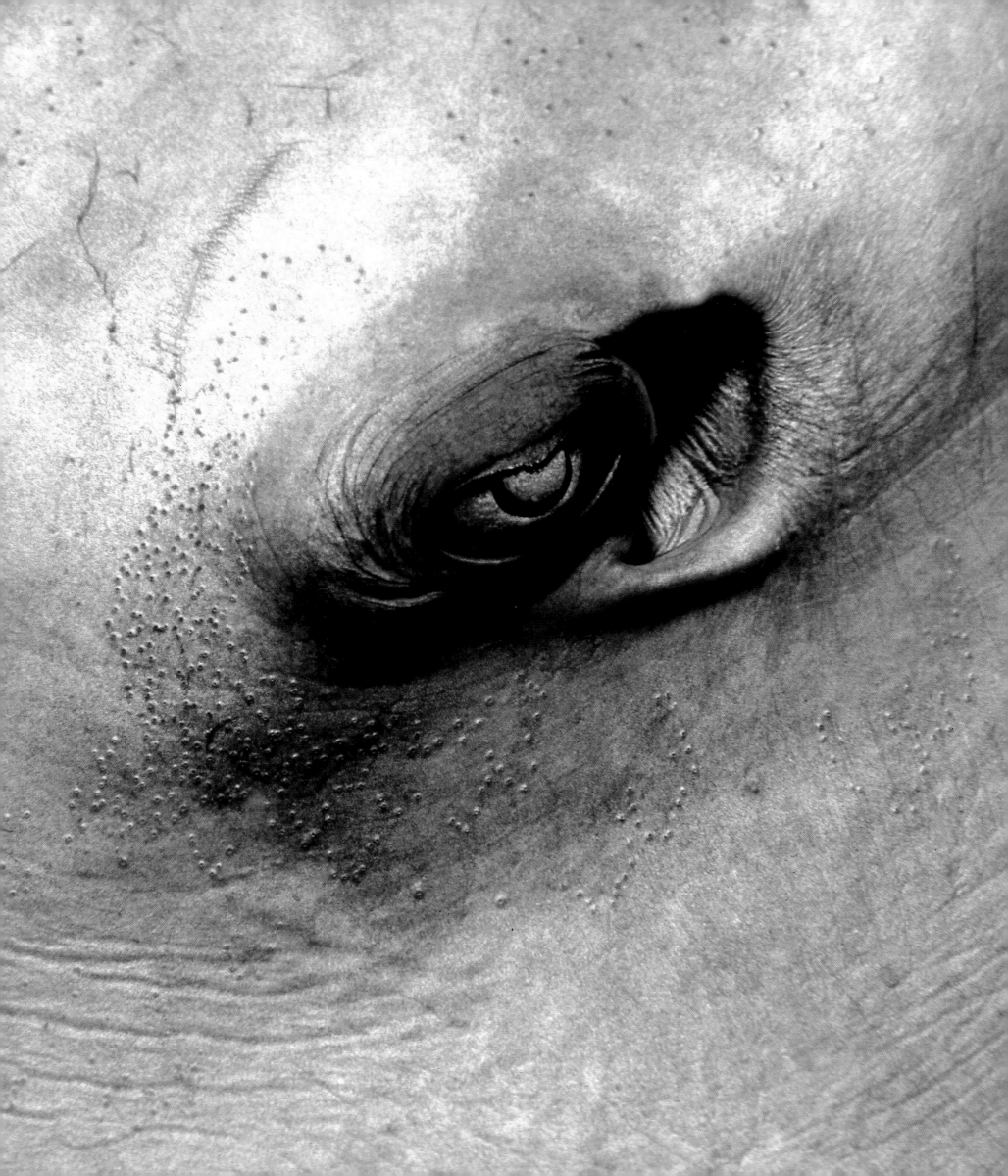

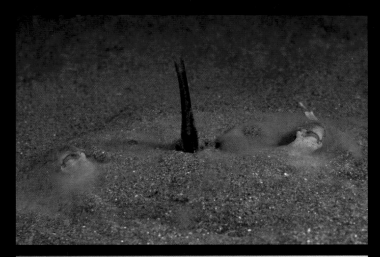

Masked crab (2 inches), France.

To get oxygen, marine animals must maintain a constant flow of water over their gills. This is not easy if you live in the sand. The masked crab makes a snorkel by putting its two antennae together like a spout. For its part, the ray breathes through an orifice located behind its eyes, using its mouth, on its underside, only for feeding.

Common Stingray
(24 in), France.

The Galley Slaves
of the Plain

Southern goatfish (10 inches), Australia.
The mound of a sea worm (8 inches), Egypt.

On the moving seafloor, an incessant rain of plankton and
detritus builds up. Buried in the sand, digging vast underground
tunnels, many small animals rummage through the sediment
in search of this food. On the sand's surface, fearsome digging
fish track down these hardworking laborers.

Bluespotted ribbontail ray (35 inches), Australia.

Mantis shrimp (15.75 inches) and spotted sharpnose puffer (3 inches), Papua New Guinea.

The Fastest Attack in All the Animal Kingdom

Mantis shrimp (15.75 inches) and the **spotted sharpnose puffer** (3 inches), Papua New Guinea.

Lying in wait, well camouflaged in its burrow, the mantis shrimp hunts, allowing only
its faceted eyes to jut out. A small pufferfish passing by triggers a dazzling attack. In a flash, the mantis shrimp
extends its plundering claws, harpoons its prey, and slowly retreats back into its burrow to devour it.
A hundreth of a second is all the mantis needs to attack its victims. This little
carnivore is the fastest predator in the world.

Sustainable Fishing: The Great Challenge

By-Catch

Fishing can result in catches that are not anticipated or wanted. A lot of fishing gear is indiscriminate and brings in dolphins, sea turtles, seabirds, non-target fish, and invertebrates, whether they are protected or not. It is difficult to estimate the loss of biodiversity through by-catch or through "ghost" fishing, where discarded or lost gill nets or traps may carry on catching fish and shellfish.

The by-catch may include the young of commercially valuable species. The small fish may have little monetary value and it may be illegal to land them if they are below the legal landing size. They are often discarded, with few surviving. Fish of legal size may also be discarded if the fisherman has no quota for them, or if the quota has already been caught. "High grading" may also take place, where the fisherman may discard fish caught earlier if fresher, more valuable fish are brought on board. In some fisheries up to 80 percent of the catch may be thrown away. Discarding is wasteful and although it may feed seabirds and other marine scavengers it can reduce significantly the numbers of fish that can grow to maturity.

Damage can be reduced by the choice of more selective fishing gears, which allow by-catch species to escape without damage. Large open meshes at the back of fishing nets may allow small fish to escape; exclusion devices may allow sea turtles to go free; the vulnerability of dolphins to surface drift nets may be reduced through the use of "pingers" and other deterrents.

Areas can be closed to fishing if large by-catches are obtained. The use of legal sanctions does not always work well, as fisheries can be difficult to police. However, fishermen can be encouraged to adopt more selective fishing methods by granting enhanced fishing rights to those who behave responsibly.

Tony Hawkins *is a specialist in fish behavior. He is a former director of fisheries research in Scotland and continues to be active in several organizations dealing with North Sea fisheries management.*

Sustainable Fisheries: The Professionals' Point of View

Fishermen are the first to be concerned about the conservation of resources, because it not only involves the condition of ecosystems, but also their livelihood. Sustainable fisheries make it possible to obtain the best possible long-term benefits from the ecosystem as a whole, while taking into account both the biological aspects of the different species as well as the economic and social aspects related to this activity. The ideal situation is the one in which the best catches can be made continually.

Sustainable fisheries are essential for businesses to be economically viable, generating stable employment while keeping "alive" the coastal areas that depend on fishing activity. This is also the only way to maintain a sustainable supply of marine protein for a population in exponential growth and that is increasingly in search of food products.

According to the FAO, 76 percent of fish stocks are in good condition: either under-exploited (3 percent), or moderately exploited (21 percent), or fully exploited (52 percent). However, 24 percent are over-exploited, exhausted or in the process of being restored (16 percent, 7 percent and 1 percent respectively). It is obvious that the situation can be improved and that it is unacceptable to have 24 percent of fish stocks in poor condition. But it is difficult to understand how the legal fishing industry is not sustainable, given the large number of rules governing it. From the point of view of the professionals, it is because the fishermen are not being involved in the decision-making process from the beginning. In order for a management system to be effective, fishermen must be convinced that the measures adopted are necessary and effective. Moreover, they must be as simple and as applicable as possible. In order to give fishermen a sense of their responsibility and actively involve them in the daily management of the fishing industry, permanent institutional frameworks must allow scientists and professionals to meet in order to carry out joint evaluations and find management methods by consensus, while taking into account the scientific data about the condition of the principal stocks. The administrations could then effectively manage fishing resources.

Javier Garat, *jurist, works in Madrid as general secretary of the Spanish Federation of Fishing Organizations (FEOPE). He is also vice president of Europêche. He collaborates regularly with the IUCN on matters related to industrial fisheries.*

Fish parks in Polynesia, once in pens patiently worked from coral, today are enclosed by galvanized wire meshes. Easy to install, these grills are being set up everywhere and are terribly effective, resulting in catastrophic overfishing on the shores of the lagoons.

Deep-Sea Fishing

Recent scientific investigations have served to confirm and reveal the truly remarkable extent of the mystery and diversity of life in the deep sea. Current estimates put the number of species inhabiting the oceans at depths below 650 feet as high as 10 to 100 million species. Many new and "relic" species (species previously known only from fossil records) have been discovered on seamounts—the peaks of underwater mountains and mountain chains found throughout the Atlantic, Pacific, and Indian Oceans. Many, if not most, of the estimated 100,000 or more seamounts worldwide may well be unique islands of biodiversity in the deep sea.

Unfortunately, the ability to reach deep into the ocean in search of new forms of life is not restricted to scientific research alone. As coastal and open ocean species of fish such as cod and tunas are overexploited, the fishing industry has developed technology to fish the ocean bottom at depths of 1.25 miles or more.

The preferred method of deep-sea fishing is "bottom trawling," which drags heavy steel plates, cables and nets across the ocean floor, destroying corals and other species that form the basic structure of deep-sea ecosystems. The scale of the threat to the marine biodiversity of the deep sea as a result is yet unknown, but potentially comparable to the threat to terrestrial biodiversity associated with the loss of tropical rainforests. Many thousands of species may be at risk, most of which are still unknown to science.

The United Nations General Assembly has responded to the concerns raised by scientists and conservation organizations, including the IUCN, and called for urgent action to protect deep-water ecosystems, particularly on the high seas. Our hope is that the world's fishing nations, and the fishing and seafood industries as well, will heed the call for the benefit of the Earth, future generations, and all humankind.

Matthew Gianni is a former fisherman who has become an internationally recognized expert on fisheries and marine conservation. He is an advisor to the IUCN Global Marine Programme and has represented the IUCN at various meetings of the United Nations and other forums.

The Ecosystem Approach to Fisheries

As long ago as 1992, at the Earth Summit in Rio de Janeiro, it was recognized that traditional approaches to natural resource management would be insufficient to achieve future conservation and sustainable development objectives. This was, in part, due to the inability of existing strategies to manage ecosystems in their entirety, focusing rather on single species or single ecosystem components. Nowadays, the need for more holistic approaches to management is widely accepted, and the so-called ecosystem approach has started to appear in initiatives and agreements.

While there are clear commitments for implementing this approach, parties to these agreements have a significant ways to go in meeting these obligations and delivering *truly* sustainable development. A comprehensive ecosystem approach to fisheries therefore requires managers to consider interactions such as:

- Predator-prey relationships
- The effects of weather and climate on fisheries
- The role of marine habitats in maintaining fisheries' productivity

Trawling is perhaps one of the most devastating of all the fishing techniques: it is indiscriminate (25 percent of the catch is thrown back into the sea dead) and causes an irreversible, machine-made erosion of the ocean floor.

- The effects of fishing on target and non-target species, as well as on marine habitats.

This approach recognizes that humans are an integral component of many ecosystems. Its overall aim is to ensure that the capacity of marine ecosystems to produce food, revenues, employment and, more generally, other essential services and livelihoods is maintained indefinitely for the benefit of present and future generations. The equitable distribution of these benefits is also a key priority.

Traditionally, fisheries management has tended to focus on individual stocks or species. Most often the data to support a more comprehensive strategy is not available or reliable and, in reality, may never be. Thus, an objective such as "maintaining a healthy ecosystem" cannot necessarily be translated into a harvest strategy and management action as yet. An initial and more practical alternative is for fisheries managers to consider how the harvesting of one species might impact upon other species in the ecosystem. Implementing the ecosystem approach in this way is much simpler in waters under national, rather than multinational, jurisdiction.

In Europe, for example, more specific and stronger measures will be needed for this approach to succeed. A key part of the solution lies in reducing fishing efforts to within a range that ecosystems can support. Innovative technical and operational changes will also be required to mitigate specific impacts on vulnerable habitats and non-target species. Such actions must now prevail over traditional institutionalized arrangements, and the industry must be given meaningful roles in developing solutions for long-term sustainability of fisheries resources if a true sense of stewardship is to be born and nurtured into the future. Relatively few fisheries currently realize their potential benefits to fishermen and society: the long-term productivity of marine systems now needs to be prioritized.

Given the state of many key stocks globally, time is not on our side. Unless radical action is now taken, we will leave future generations with a degraded legacy, as well as many questions about the current generation's failings to provide adequate protection when it was needed the most.

Dan Laffoley is head of marine conservation at English Nature, the government agency for nature protection in the United Kingdom. He is a specialist in marine ecology and thematic team leader for the World Commission on Protected Areas (WCPA) Marine Biome.

Olive rockfish (24 inches) and black rockfish (22 inches) in bull kelp forest (33 feet), Canada.

The Undersea Forests

Forests are not restricted to land. In cool temperate areas, the shallow, rocky seafloor is colonized over vast areas by giant seaweed: the laminaria. The differences between land forests and undersea forests clearly illustrate the specificity of life in the water.

In order to reach up toward the surface and the light, laminaria do not need rigid trunks, but simply have floaters filled with gas. Neither do they have roots, since the mineral elements essential to their growth are to be found dissolved in the sea water. As for water, which is the main factor limiting the growth of land plants, here it is available an infinite amount!

In the warm season, when the springtime sun begins to shine and provide the light energy needed for the development of the seaweed, the laminaria then grow at breathtaking speed (up to more than 20 inches a day for some species), plankton proliferates, animals have many and varied shelters, thanks to the seaweed, while the great predators, whether they are sharks or dolphins, hesitate to venture into the half-light of this labyrinth.

Copper rockfish (22 inches), **Pacific red sea urchins** (8 inches) and **Southern stiff-stiped kelp** (5 feet), Canada.

To keep the underwater forest's surface from decreasing,
the growth of the seaweed must compensate for physical destruction (storms, waves, harvests),
and consumption by herbivores. Many species participate in maintaining this balance: herbivorous sea
urchins, but also all those creatures that eat them, their parasites, all the way up to large predators, such as sea
lions. The productivity and capacity of an ecosystem to evolve in the event of change increase when
it is composed of a large number of species. This is one of the reasons scientists are so concerned
about the current problem of biodiversity erosion.

South American sea lion (9 feet) in **giant kelp forest** (115 feet tall), Chile.

Long-snouted seahorse
(5.5 inches), France.

An Undersea Jungle

Blue maomaos (16 inches), New Zealand.

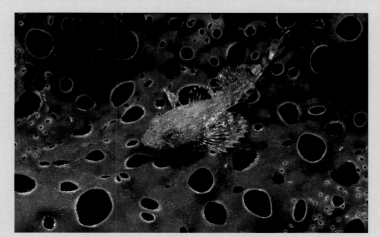

Arctic sculpin (5 inches), Canadian Arctic.

In the maze of seaweed, animals try to find something
to eat without being eaten by a more powerful predator.
The plankton-eating blue maomaos stay close
to the forest, where the plankton is more abundant,
and then take refuge in it at the first sign of danger.

Brown kelp (8 feet), Chile.

Sea rose algae and **Grotto goby**
(3/4 inch), France.

Forkweed (14 inches), **Zebra seabream** (16 inches), **Mediterranean rainbow wrasse** (6 inches),
White gorgonians (20 inches), France.

A seaweed has no stems, leaves, or roots.
All of these sophisticated parts are the privilege of the so-called higher plants.
It manages quite well without them. From the surface to nearly 300 feet down,
from the tropics to the poles, in freshwater or saltwater, seaweed has been present
on Earth for about nine hundred million years. But not all seaweeds are the same.
In a simplified way, they can be broken down into a green linage, a red linage, and a
brown linage. And though they may look alike, there are more differences between
red seaweed and green seaweed than between green seaweed and a rosebush.

Gilthead seabream (10 inches),
Dead man's fingers (8 inches),
Peacock blenny (6 inches),
Moon jelly (8 inches),
Red ascidian (12 inches), France.

Giant kelp
(100 feet tall), Chile.

The bladder of bull kelp (4 inches in diameter), Canada.

Yellow mussels in kelp (3/4 inch), Chile.

Seaweed, like all plants, is capable of converting energy from the sun
into chemical energy and making it available in the form of sugars.
To achieve this, seaweeds need light, which is why some types have
floats filled with gas that allow them to hoist their fronds toward the surface,
and which also compensate for the weight of the many organisms that
grow on and weigh down their "leaves."

Rising With the Help of Buoys

Leafy seadragon [18 inches], Australia.

The Leafy Seadragon

Leafy seadragon (18 inches), Australia.

Fish have been evolving and becoming differentiated in the oceans for 500 million years.
So it is not surprising that some of them now look especially strange. The leafy seadragon is one of these odd
creatures. Its seaweedlike appendages—the result of evolution, which causes changes by random chance and is guided
by the survival of the fittest—allow it to be perfectly camouflaged in the undersea forests of Australia.

Copper rockfish (22 inches) in a forest of bull kelp (33 feet), Canada

The Crab That Hides the Forest

False king crab (3.75 inches), Chile.

Providing shelter, a direct source of food, hunting grounds, and a means
for staying in place, undersea forests are essential to the survival of thousands
of animal species. As long as there is enough light, it is the plants that
make the landscape, not the animals. However, more often than not, we
prefer to notice the unusual appearance of the false king crab crab or the grace of
the old wives. The plants seem to make up the background scenery, and the
animals hide the forest—how unfair!

Old wives
(10 inches), Australia.

Saltwater crocodile (10 feet), Papua New Guinea.

Mangrove forest at low tide,
Europa Island, Mozambique Channel.

Humidity, the nauseating smell of stagnant water, mosquitoes, a slippery and wet
seafloor where the foot sinks, sometimes even a crocodile—here is an ecosystem
that looks quite inhospitable at first glance. However, the mangrove swamp provides many services: it is a nursery for young fish, a barrier against the
destructive forces of the ocean, a buffer zone between the sea and the land.
And when the sea's clear water sweeps in at high tide onto the interlacing roots of the
mangrove trees, it gives a completely different appearance to the mangrove swamp.

Dory snapper (8 inches) among the prop roots of Asiatic mangrove, Europa Island, Moza

e Channel.

The Secret Adolescence
of the Green Sea Turtles

As soon as they hatch, green sea turtles dash toward the open sea, where there are fewer predators.
Ten or so years later, they return to mate and lay their eggs on the same site where they were hatched.
Little is known about the places they frequent during their first years of life. While it was believed
that they spent that time in the open sea, scientists recently observed a large number of
young turtles in the mangrove swamp of Europa Island, Mozambique. Where did these turtles come from
and why do they gather together in such a strange place? It remains a mystery.

Juvenile green sea turtles (14 inches), Europa Island, Mozambique Channel.

Mangrove swamp at maximum ebb tide, Mozambique Channel.

Forest of mangrove trees at high tide, Mozambique Channel.

Marine Protected Areas:
The Management Tool for the Oceans of the Future

Why establish Marine Protected Areas (MPAs) and organize them into a network? The short answer is to prevent degradation of both biological diversity and ocean productivity. Indeed, traditional management methods have been a failure in most of the world's sea areas. The United Nations Food and Agriculture Organization (FAO) estimated in 1995 that 69 percent of the world's marine fisheries were "either fully to heavily exploited, overexploited, or depleted...and therefore in need of urgent conservation and management measures." Events since 1995 have shown that this trend is continuing.

Traditional management methods

Many scientists believe that the primary cause of such failures is inherent uncertainty. The development of chaos theory by Lorenz and, independently, by May shows that cause-effect relationships that contain non-linear elements are likely to be characterized by dramatically different outcomes from small changes in initial conditions. The behavior of marine ecosystems over time is indeed often non-linear. It follows that management based on linear thinking—unlike MPAs—is likely to fail often. It is clear from the fundamental nature of these problems that they are unlikely to be solved merely by expenditure of more effort and resources on traditional ocean management methods. These methods need to be supplemented by other approaches that are fundamentally different. MPA networks are seen by many to be the appropriate response, because they protect habitat and ecological processes, even when those processes are not fully understood.

Marine protected areas

The term "Marine Protected Area" is defined by IUCN as: "Any area of intertidal[1] or subtidal[2] terrain, together with its overlying water and associated flora, fauna, historical, and cultural features, which has been reserved by law or other effective means to protect part or all of the enclosed environment."

The primary reasons for creating MPAs:

- to maintain essential ecological processes and life support systems;
- to ensure the sustainable utilization of species and ecosystems;
- to preserve biological diversity

The major problems that stand in the way of achieving these aims stem from stress from pollution; degradation and depletion of resources, including species; conflicting uses of resources; and damage and destruction of habitat. MPAs, in association with other management methods, can address all these problems within an ecosystem. Together, they can establish an integrated management system providing levels of protection varying throughout the area, from total exclusion of human activity other than research and monitoring in relatively small areas, to usually larger areas where many non-destructive activities are allowed or encouraged. Ideally, this integration should extend to coordinated management of marine and terrestrial areas in the coastal zone and beyond, with special emphasis on controlling land-based sources of marine pollution. Experience has shown that the presence of a downstream MPA can be a powerful argument for changing polluting land activities.

Conclusion

This approach provides for the protection of habitat. This is considered essential if the three major aims described above are to be achieved, because the interdependencies and variabilities in most ecosystems are unlikely to be so simple as to allow protection merely by management focusing on particular species or human activities, often called sectoral management.

The World Summit on Sustainable Development (WSSD) in 2002 set a target for the establishment of a global representative system of MPAs by 2012. This aim has been endorsed in various fora, including the United Nations, the World Parks Congress, the World Conservation Congress in 2004, and the Conference of Parties of the United Nations Convention on Biological Diversity.

The most recent evaluation of the existing global system of MPAs, together with priorities for new MPAs was carried out in 1994 by a coalition of the World Bank, IUCN, and the Great Barrier Reef Marine Park Authority. An update of that publication is urgently needed. The most striking development associated with this aim has been the recent rapid increase in support for a representative network of MPAs on the high seas, outside national jurisdiction. IUCN has been at the forefront of this development, in collaboration with NGOs, such as the Worldwide Fund for Nature (WWF), and various governments. IUCN's World Commission on Protected Areas (WCPA) has established a formal Task Force, which is doing everything that it can to catalyze and accelerate this initiative.

Graeme Kelleher has worked on natural resource management for the last 30 years, in Australia and worldwide. He was president of the Great Barrier Reef Marine Park for 16 years and also served as vice-president of the IUCN World Commission on Protected Areas, Marine Program. He is now an active member of the High Seas MPA Task Force of IUCN.

1 Intertidal: area located between the elevation of the lowest yearly tide and the elevation of the highest yearly tide.
2 Subtidal: area below the low-tide level.

While 12 percent of the land has been set aside for conservation in more than 100,000 protected areas around the world, the sea has only 1,350 marine protected areas, accounting for barely 0.5 percent of the oceans' surface.

THE MARINE PROTECTED AREAS IN THE MEDITERRANEAN

The creation of marine protected areas (MPAs) was only begun in the Mediterranean in the early 1960s, and of the 60 current MPAs, only 14 were created before 1980. The total marine area covered by these MPAs is about 1.25 million acres, most of which are located on the north shore of the Mediterranean Basin, each averaging about 12,355 acres. While some of these MPAs are managed well enough to play an essential role in the preservation of their natural marine heritage, even in some cases an important role in the economic development of their regions, most of them suffer from inadequate means, planning and system of governance.

The systems of governance of the Mediterranean MPAs are currently based in large part on public authorities (state or local), and the participation of the NGOs and the private sector is still very limited.

The Mediterranean was the first region in the world to be granted an international agreement specific to marine protected areas and coasts (the SPA Protocol of the Barcelona Convention). This indicates the willingness of the countries in the region to develop their cooperation in this field. Moreover, this protocol was further developed in 1995, particularly with the creation of a list of Specially Protected Areas of Mediterranean Interest (SPAMI) that can even be located outside areas under national jurisdiction. The registration of a site on the SPAMI list entails recognition by all parties to the Protocol, who thus undertake to abide by the protective and management measures planned for that site. Much hope has been invested in the SPAMI system for improving the management of the Mediterranean MPAs and for assisting the region to have a representative MPA network.

Chedly Rais was a researcher at the National Oceanographic and Fishing Scientific and Technical Institute of Tunisia. The former scientific director of the United Nations Environment Program / Mediterranean Action Plan Regional Activity Centre for Specially Protected Areas, he now works for large regional and international agencies like the IUCN.

Marine protected areas allow fish stocks to regenerate by shielding them from being fished, but these areas remain permeable to all types of pollution. Thus, they are not spared from litter resulting from objects dumped into the sea; effluents from ships and factories; and discharges from wastewater treatment plants.

AN INTERNATIONAL MPA-NETWORK FOR NORTHERN EUROPE

A large number of international forums have called on nations to establish a worldwide network of marine protected areas by 2012. Against this background, the Northern European Ministers of Environment and the member states of both the Oslo/Paris (OSPAR) and Helsinki (HELCOM) conventions jointly agreed in 2003 to develop a network of well-managed marine and terrestrial protected areas (MPAs) by 2010. This network will cover the Baltic Sea area and the OSPAR maritime area, reaching from the mid-Atlantic ridge and eastern Greenland to the European coasts, and from the North Pole to the Azores. In addition, the European Community required all of its member states to also include important sites from their coastal and Exclusive Econimic Zone waters in the NATURA 2000 network of protected areas by 2008.

On the ground, the OSPAR commission, the HELCOM commission, and the European Community are working together to achieve a network of MPAs for the 18 nations concerned. Whereas HELCOM has already established 62 MPAs since 1994 with another set of about 20 areas under consideration, OSPAR only started working on MPAs in 1999. The OSPAR maritime area includes large parts of the high seas (i.e., the ocean beyond national jurisdiction), which makes the establishment of MPAs particularly challenging and needs clear support from the United Nations.

The establishment of the NATURA 2000 network is legally binding for all its member states, but it comprises almost solely terrestrial areas. Only a few countries have so far submitted proposals to the European Commission for the inclusion of MPAs in the NATURA 2000 network.

*The **OSPAR** and **HELCOM** Commissions have elaborated a set of tools to facilitate the establishment of MPAs. They include: selection criteria for the creation of MPAs; identification of management needs; questions of stakeholder participation; involvement of relevant international bodies; ecological quality objectives, a Red Data book of biotopes, and initial lists of threatened and/or declining species and habitats that should benefit from MPA protection.*

***Henning von Nordheim** works at the German Federal Agency for the Conservation of Nature as head of the Coastal and Marine Conservation Department. He also leads the OSPAR program on protected areas, species, and habitats. He is an active member of the IUCN World Commission on Protected Areas.*

Whip coral (5 feet), Safaga, Egypt.

The Undersea Mountains

There is an unnamed landscape in the oceans. There is no word to describe it—it remains to be coined. On land, natural landscapes such as forests, savannahs, and prairies are nearly always characterized by their vegetation. The trees and plants form a stable setting in which the animals evolve, always hurried, often invisible, busy with finding food or escaping from a predator.

This is not the case in the oceans. On the rocky ocean floor, there are cliffs and fallen rocks, especially starting at a depth of about 65 feet, where darkness takes over and prevents the growth of seaweed. This rocky floor is covered with a multitude of animals that look like plants crowding each other. What should this very special habitat be called, this forest of unmoving animals? Gorgons, sponges, anemones, none of these animals have legs or eyes and they remain completely immobile all their lives. Although outside the water a rock is practically devoid of any colonization, in the underwater world it is one of the richest habitats, providing a solid base to cling to and find protection from the undertow. In calm or choppy areas, in shadowy or bright places, and offering multiple hideouts in case of danger, here every creature can find a suitable home. It is the complexity of this habitat that explains the extraordinary diversity of its inhabitants.

The Paradise of Attached Animals

Mediterranean feather star (6 inches in diameter), France, and **European fan worm** (12 inches).

On the boulders exposed to the ocean currents
but plunged into the shadows, plants
are replaced by animals that live attached in one place
and feed by filtering the seawater.
Each has its own technique: the gorgons have
hunter polyps, armed with tiny urticant tentacles,
while the European fan worm and the feather star
deploy branched organs that act like cobwebs.

Gorgonian coral of Vancouver (12 inches), Canada.

Moontail bullseye (16 inches), Egypt.

For many animals, having a refuge is essential.
From tiny cracks to vast grottoes, safe havens are many and varied in the rocky ocean floor,
and their diversity explains in large part the wealth of their fauna.

Black rockfish (24 inches), in **bull kelp** (33 feet), Canada.

Royal paper bubble (1 inch), French Polynesia.

Fish in the Jacuzzi

The White Island volcano, New Zealand.

Forests burned by acid rain from the White Island volcano.

Sandager's wrasse (18 inches) and the **Red pigfish** (18 inches), New Zealand.

White Island, off New Zealand, is an active volcano that spews out gases so toxic
that they burn the forest when they fall onto the island in the form of acid rain.
This harmful effect is not noticeable underwater and
does not hinder the development of abundant flora and fauna.

European spiny lobster (16 inches) and **Red seasquirt** (4 inches) on a coralline bottom, France.

Many attached animals, as well as some kinds of red seaweed, secrete calcium to make their tubes, shells, or hard blades.
Down deep, where the light is dim, these organisms can completely cover the rocks. Over time,
their calcified parts accumulate and reach a thickness of several feet, producing a "living rock" called coralline.

Dusky grouper (3 feet) and **White seabream** (12 inches), Spain.

Decorated warbonnet (12 inches), Canada.

The wreck of the *Cedar Pride*, 260-foot cargo vessel, Jordan.

For the billions of animal larvae that drift in the ocean, finding a support to attach to or a shelter where they can hide is a matter of life or death. As long as their cargoes do not include toxic products, shipwrecks are a godsend for these creatures. Offering a solid, free-standing support with plenty of hiding places in which to take refuge in case of danger, a sunken ship is not a loss for everyone.

Orangelined cardinalfish (5 inches), **the wreck of the _Cedar Pride_**, Jordan.

Fjordland, New Zealand.

Bottlenose dolphins
(13 feet), New Zealand.

Black coral
(10 feet), New Zealand.

When the glaciers withdrew from New Zealand ten thousand years ago,
they left deep gashes in the mountains that were overrun by the sea.
In these fjords, the rains are heavy and flow down the steep slopes.
They are transformed into a yellow tealike liquid, rich in plant tannins,
which is deposited delicately on the saltwater without becoming mixed with it.
The overlay of this yellow water with the blue seawater creates a unique
green environment that prevents the growth of seaweed and allows attached animals
to blossom a few feet from the surface.

The Fjords

The Red Gold of the Mediterranean

Albino red coral (4 inches), France.

Red coral is one of the symbols of the Mediterranean.
Eight thousand years ago Neolithic people were already adorning themselves with it, and it was so important
that red coral figures in Greek mythology. According to legend, the drops of blood falling from the head of the
Gorgon decapitated by Perseus to save Andromeda were petrified and gave rise to the red coral that spread
through the oceans. History, however, does not say how the extremely rare albino coral was formed. While the
albino coral is of no interest to jewelers, it is a delight to naturalists.

Red coral
(8 inches), France.

Lingcod (5 feet), Canada.

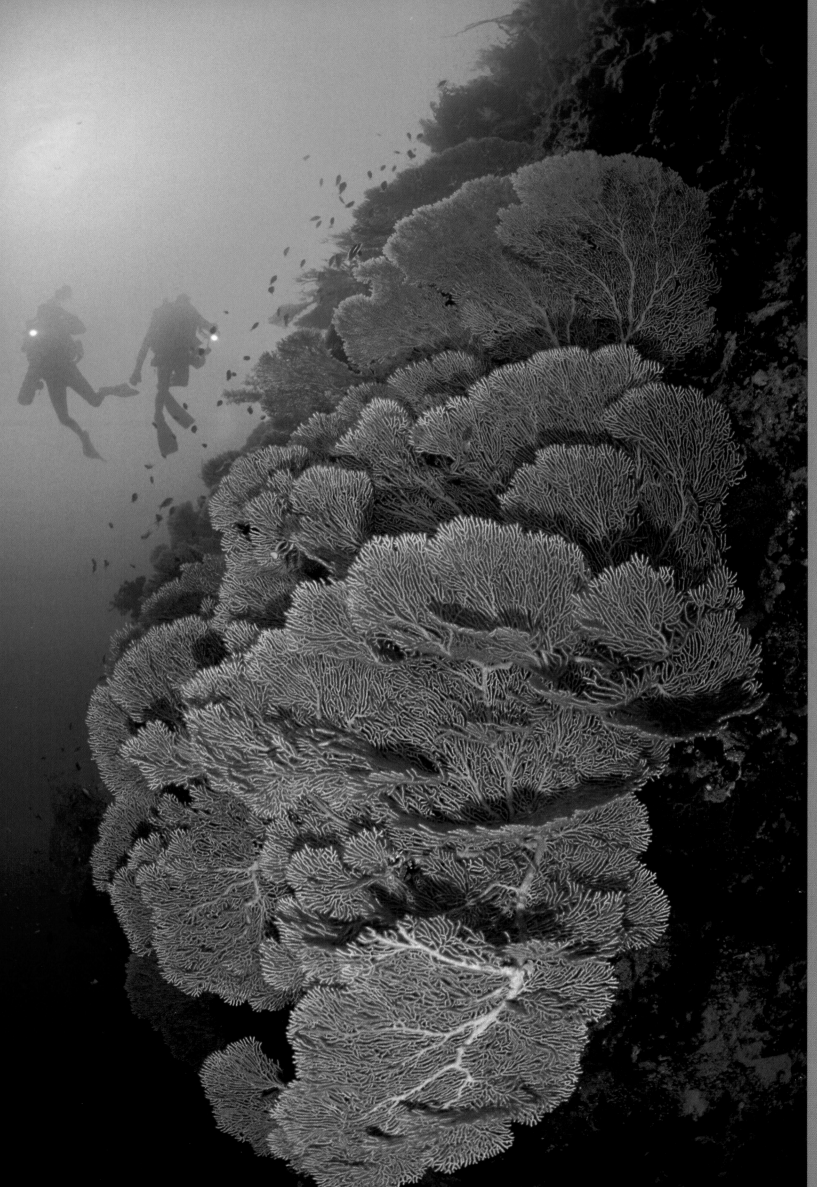

Giant gorgonians
(10 feet), Egypt.

Attached Life, Fragile Life

For attached animals,
it is impossible to flee if the
environment becomes unsuitable
for their survival. They are,
therefore, unquestionable
indicators of the quality of the
water. Dense populations of
giant gorgonians in the Red Sea
indicate a constant environment
with no violent changes, while
just below the surface of
the New Zealand fjords, the bare
walls bear witness to waters
subject to violent disturbances
(here, sudden natural desalting)
unsuitable for their survival.

Eleven-armed sea star
(20 inches), New Zealand.

Sea Turtles: A Marine Icon

The seven species of sea turtle found throughout the world's tropical and temperate oceans are among the most maligned and endangered, yet poorly understood creatures of the sea. They are the only fully marine-adapted members of the order Testudines (turtles, tortoises, and terrapins) and include seven species—the green turtle, the flatback, the hawksbill, Kemp's ridley, the olive ridley, the loggerhead, and the leatherback.

First appearing some 110 million years ago, sea turtles represent a deeply rooted and irreplaceable component of ocean ecology. Sea turtles witnessed, but were unscathed by, the dramatic events that left the dinosaurs extinct, and have weathered, nearly unchanged, the ensuing radiation of most of life on Earth, including humans. Yet today, the continued existence of sea turtles hangs in the balance—precariously close to extinction —as they meet the greatest challenge in their evolution: post-industrial man.

Peculiar creatures

Sea turtles are found nesting on beaches in all tropic regions; foraging in near-shore habitats, seagrass pastures, and coral reefs; and traversing the open ocean, from surface to abyss. They exhibit a plethora of unique and unusual natural history traits.

The leatherback turtle, for instance, is one of the most extraordinary of Earth's creatures. To begin with, it is the largest of all extant Chelonians, with specimens on record measuring more than 10 feet in length and weighing over 2,000 pounds. Furthermore, it achieves this enormity on an exclusive diet of jellyfish, an organism that itself is mostly water! Leatherbacks can make several dives each day to depths of over 3,000 feet, where the pressure, temperature, and darkness preclude all but the most highly specialized life forms; and annually they traverse thousands of miles of open sea, crossing entire oceans sometimes multiple times.

Endangered

All sea turtle species are considered endangered or critically endangered by the *IUCN Red List*, with the exception of the flatback, whose status is yet unknown. Some populations, like the leatherback in the eastern Pacific, are drastically reduced as a result of years of uncontrolled egg collection and a precipitous increase in the incidental capture by different fisheries. The population of leatherbacks nesting at Playa Grande, Costa Rica, has dwindled by more than 90 percent in just a decade. This followed similar reports from researchers in Pacific Mexico, who witnessed an abrupt population decline in the nesting colony there: from an estimated 70,000 leatherbacks in 1982 to fewer than 250 in 1998–1999.

Luxury market threat

The threats to sea turtles worldwide are numerous: habitat alteration and destruction; hunting and egg harvesting; and incidental capture in fisheries. Sea turtles have long been a source of protein for coastal peoples, and direct take by humans remains a substantial threat in many areas. Turtles have also fed luxury markets, including a formerly heavy trade in green turtle calipee (fat) for turtle soup in Europe and the United States, a taste that has now mostly disappeared. Trade in tortoiseshell, the product that derives from the scutes of the hawksbill turtle, also remains a problem despite the prohibition of international trade of sea turtle products.

Marine pollution

More modern hazards to the sea turtle include marine pollution (especially ingestible plastics), industrial fishing, beach lighting that disturbs their nesting and disorients hatchlings, and even diseases such as *papillomatosis*, a potentially lethal virus never seen in sea turtles before the 1960s.

Getting our house in order

As go sea turtles, so go the oceans. And as go the oceans, so goes man. Sea turtles today face greater challenges to their survival than ever before in the past 110 millennia. Yet even greater are the challenges faced by man to change the way we act, and the role we play within the biosphere that is our common home. In the case of sea turtles, their survival can be ensured by a few simple changes in human behavior:

- improving the ways we fish—using technology to minimize the adverse impacts of longline fishing, shrimp trawling, and gill netting;

- taking greater care in the way we develop our coastlines, leaving habitats intact and avoiding light pollution; and

- being aware of the impacts of stray plastics, industrial runoff and other pollutants.

Humans are a remarkable species—we have traveled to space, walked on the moon and returned home alive, and described the cosmos and matter itself down to its subatomic units. It is not beyond us to be conscious of our place in nature and to find ways to make our life on Earth sustainable. Sea turtles are here to remind us that our own survival depends on it.

Brian Hutchinson works for Conservation International's Sea Turtle Flagship Program. Rod Salm also works for Conservation International, working to protect critical ecosystems in tropical Latin America and the Caribbean, Africa, Madagascar and Asia. Both of them help manage the IUCN-SSC Marine Turtle Specialist Group.

Following the explosive growth among human populations, turtle fishing, even when carried out traditionally, has been prohibited everywhere in the world to avoid the major risk that some species of turtles might disappear. But traditions have a persistent life, and the measures to control this fishing remain far too few.

Hawksbill sea turtle (3 feet), Polynesia. The hawksbill can be coveted only for its shell, because its flesh is highly toxic, a fact that has, in part, protected this species from being poached.

WHAT IS THE FUTURE OF SEA TURTLES IN THE MEDITERRANEAN?

Among the seven species of sea turtles, five have been observed in the Mediterranean: the loggerhead turtle, the green turtle, the leatherback turtle, the Kemp's Ridley turtle, and the hawksbill turtle. The loggerhead turtle and the green turtle reproduce in the Mediterranean and, with the leatherback turtle, are the most common turtles in all of the Mediterranean basins where the young as well as the large adults can be observed. In the western Mediterranean, there are several thousands of turtles in migration from the western Atlantic, in addition to the native specimens.

The loggerhead turtle, *Caretta caretta*, is the most abundant species, ranging from the Strait of Gibraltar to Lebanon; its egg-laying beaches are in Greece on Zakynthos Island, in Turkey, Libya, Israel, Tunisia, and Cyprus. Some egg-laying has been observed in Egypt,

on the Italian islands, and once on the Spanish coast in 2001. On Zakynthos, the number of nestings can reach 2,000 per year. The Libyan and Tunisian coasts are the most important egg-laying areas along the southern shores.

The Mediterranean population of green turtles, *Chelonia mydas*, is in danger. The exploitation of this species from 1930 to 1983 led to a decline in Mediterranean reproductive stocks. Today, the species is still reproducing in Turkey, Lebanon, Israel, Egypt and Cyprus. But its situation is becoming worse because of the loss of egg-laying beaches, the deterioration of marine plants and seaweed habitats, pollution and increasing pressure from fishing.

Boats equipped with surface trawl lines to fish for swordfish, bluefin tuna, and albacore also pull in thousands of loggerhead turtles each year, putting the species at serious risk. Green turtles do not fare much better, when we take into account catches by trawlers, driftnets and all of the local fishing methods along the Mediterranean coasts. Then there are also the slower

deaths caused by injuries from fish hooks, lines, driftnets, as well as those caused by "ghost" catches.

A study of two turtle populations in the Mediterranean shows that mortality caused by fishing and other human activities could affect not only native but also foreign populations. The efforts to preserve these populations on their beaches of origin could be seriously compromised if the main causes of mortality are not reduced in their habitats, including on the migratory routes, in the Mediterranean.

Juan Antonio Caminas is director of the Malaga Center of the Spanish Oceanographic Institute. A biologist specializing in fishing, he has been interested in sea turtles for more than 20 years and participates in the work of the Marine Turtle Specialist Group of the IUCN.

The Oases of the Open Ocean

The dunes of the Sahara are a lush garden compared to the deserts of the open ocean. Away from the coasts of the great oceans, the water becomes pure and life becomes extremely rare. The oceans, as monotonous as they may seem, are far from being homogenous. They are composed of a mosaic of different waters, more or less rich, which mix poorly. Thus, there are pockets of fertile waters: often these are cold waters arriving from the poles thanks to powerful currents, but they can sometimes also be deep waters propelled from the depths by obstacles on the bottom acting as natural trampolines.

In these oases of the open ocean, the water seems miraculously to become fertile again. Life explodes, while the great predators—dolphins, sharks, and whales—appear suddenly out of nowhere to take part in the feast.

Without shelter where they can rest, with no ground to settle on, the inhabitants of the open ocean are condemned to perpetual wandering. But more than anything, they are faced with a considerable problem: how to not sink! To meet this constant challenge, the inhabitants of the open ocean have developed amazing adaptations. From the blue whale, the largest animal the Earth has ever known, to the tiny copepods, which look like something out of science fiction, this is where the most fascinating creatures of our planet live.

Red phytoplankton (0.0015 inch), France. Colorized image from a photograph taken by electron microscope.

The microscopic algae that make up the phytoplankton are ten times smaller than a grain of fine sand. Like all plants, they need light, but they also need mineral elements, which, in the irony of marine life, are primarily confined to the depths. Here and there, owing to rising currents, the mineral elements are propelled toward the surface. The phytoplankton then multiply very rapidly, spurring a huge growth of the zooplankton, and thereby putting into place the whole oceanic food chain up to the largest animals.

Marine diatom (diameter: 0.0012 inch), France.
Colorized image from a photograph taken by electron microscope.

Blue shark (7 feet), France.

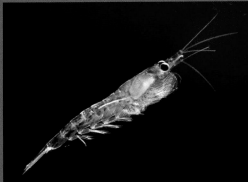

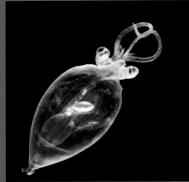

Young squid (4 inches), France.
Northern krill (1.5 inches), Mediterranean.

The larva of an octopus (1/3 inch),
Subantarctic region.

Amphipod (3/8 inch),
Subantarctic region.

A large portion of zooplankton only approach the surface at night, coming up to
feed on the phytoplankton before returning in the morning to a depth of
about 500 feet. They don't come up alone. They are followed by predators, such as
hordes of shrimp and fish larvae, which devour them and, in turn,
are hunted by more powerful predators, such as squid. These vertical migrations
are huge compared to the size of these tiny organisms, and they constitute
one of the greatest movements of animals on the planet.

Blue jellyfish
(12 inches), Sweden.

The Mountaineers of the Open Ocean

Humpback whale (20 feet), French Polynesia.

Great Spaces for Great Nomads

Basking shark (26 feet) and **moon jellies** (8 inches), Great Britain.

Sailfish, whales, sharks, or manta rays—these numerous, tireless nomads travel tens of thousands of miles to return to feeding areas, as well as to sites favorable for their reproduction. Their great size allows them to make these impressive migrations in order to connect these two necessities of life.

Indo-Pacific sailfish
(6.5 feet), French Polynesia.

The Initiatory Journey

Juvenile bluespine unicornfish (4 inches), French Polynesia.

At the beginning of its life, the bluespine unicornfish undertakes an initiatory journey within the vast oceans. Released as an egg in the depths of the reef, then subjected to severe hardships with its brothers and sisters, it will be transformed, after a journey in the open sea of several weeks, into a transparent and still awkward larva that will return to the lagoon. The development of the bluespine unicornfish will not stop there. The fish will require much more time and experience before becoming an adult. The natural selection for this species is such that less than one egg in a million will grow to have the characteristic horn of the mature individuals.

Postlarva of the bluespine unicornfish
(1 inch), French Polynesia.

Postlarva of the broadbarred firefish (1/2 inch), French Polynesia.

The Odyssey
of the Larvae

Postlarva of the leaf scorpionfish (3/4 inch), French Polynesia.

The larva of the leaf scorpionfish is full of mystery.
Its translucent body does not reveal a single red trace of
hemoglobin, unlike the larva of the broadbarred firefish,
whose gills, red with blood, are clearly visible.
But hemoglobin is the protein that allows the transport of oxygen.
So how do the leaf scorpionfish larvae get oxygen?
Scientists believe that the larvae are so thin that
simple contact with oxygen through the larvae's skin is enough
to oxygenate all of the cells in their bodies.

Adult leaf scorpionfish (3 inches), Malaysia.

Ocean Governance—Better Decisions for Better Management

When the early seafarers set off to find new fishing grounds or a more hospitable place to dwell —or simply to explore what lay beyond their immediate horizon—they could be fairly confident that unsettled areas elsewhere would yield rich marine resources. They did not have to worry about crowded sea-lanes or about how to dispose of their waste products. At some stage, however, agreements were necessary to avoid over-harvesting and the collapse of resources and to segregate waste disposal sites from human use areas. The rudiments of "ocean governance" began to emerge. As coastal communities asserted rights to particular offshore resources and denied them to others, either they had to defend those rights with potential loss of life or come to agreement with neighboring groups.

These agreements over the use and conservation of common resources, and over the rights and boundaries of countries bordering the sea, remain the two main tenets of ocean governance. These basic tenets are laid down in the United Nations Convention on the Law of the Sea (UNCLOS). In areas within national jurisdiction, responsibility to conserve and manage resources rests with the coastal state. This state also has the authority to protect the marine environment and control marine scientific research. In areas beyond national jurisdiction, since no country may claim sovereignty or exclusive rights over any part, all must cooperate to ensure conservation and marine environmental protection.

National jurisdictions

UNCLOS allows all countries to manage marine resources out to a limit of 200 nautical miles, an area called the Exclusive Economic Zone (EEZ), and in some cases beyond. Some 70 percent of the world's fish harvest comes from within the EEZ, and nearly 80 percent of marine pollution comes from activities on land. Clearly, the greatest challenges of ocean governance are within national jurisdiction. Yet granting coastal states control in these areas—a major change in ocean governance—did not avert the "tragedy of the commons" (over-exploitation due to free access and unrestricted demand)—as was expected. Few countries have a good record in managing their fisheries and maintaining healthy coastal environments. Between increasing human pressures and the political influence of powerful fishing and industrial communities, decision makers have not been pressed to make the right choices. In addition, many developing nations lack the means to manage coastal resources effectively and enforce their laws, whether at the local level or vis-à-vis irresponsible fishing fleets flying "flags of convenience" from other countries. Improvements in national management are essential, and international collaboration will be needed.

International agreements

At the level of international ocean governance, new challenges are also emerging. Growing human impacts expose more linkages between states—through transboundary fish stocks or pollution, migratory spe-

Each year, the Mediterranean, a semienclosed sea, receives 650,000 tons of hydrocarbons (the equivalent of thirty oil slicks of the size of the spill caused by the tanker Erika). To this pollution must be added the effect of wastewater. More or less treated, the wastewater carries significant quantities of toxins and other heavy metals, which accumulate in the food chain. A stranded whale was thus found to have more than 2 pounds of mercury scattered throughout its body.

Out of sight, out of mind...The high seas are far from the jurisdiction of coastal nations. So it is difficult to enforce international laws aimed at limiting the exploitation of the seas' resources.

cies, and shared ecosystems. These problems cannot be solved by nations acting alone. UNCLOS is supplemented by numerous global and regional agreements that elaborate its basic "constitution." These more detailed "ocean governance" rules address specific uses, such as shipping, fishing, or waste disposal at sea. Other goals are to protect the marine environment from pollution caused by offshore oil and gas development or borne to the sea by rivers or through airborne auto or industrial emissions. Far greater financial and technical resources are needed to help all states meet these goals. The international bodies established by agreements play a vital role in stimulating cooperation and strengthening the exchange of knowledge and technical skills. They foster effective ocean governance at national and international levels.

New challenges for ocean governance are also emerging in areas beyond national jurisdiction. This vast expanse offers major opportunities. There is still time to preserve its important reservoir of marine species and habitat—indeed, to ensure that the marine ecosystems retain their productivity and resilience (i.e., their ability to regenerate). Perhaps this will help replenish degraded coastal systems, too.

Conclusion

Can we now begin to anticipate, in light of current experience and trends, how to prevent and reverse adverse changes? Although substantial resources are devoted to assessing the risks and impacts of global climate change, the monitoring and assessment of the marine realm remains sketchy. While we know enough to act, greater knowledge can help stimulate sufficient public pressure and thus political will, especially in those vast ocean areas that are "out of sight and out of mind." As human

use of the oceans has grown more intense, effective ocean governance is increasingly a necessity.

Lee A. Kimball is a specialist in international marine law and policy and serves as an advisor to IUCN's Global Marine Programme. She has advised intergovernmental and nongovernmental organizations on the development of international ocean law and institutions since she began her career in 1974 at an NGO working on the development of the United Nations Convention on the Law of the Sea.

THE HIGH SEAS—AN AREA IN NEED OF PROTECTION

Vast expanses of the ocean lie beyond the jurisdiction of coastal nations. These "high seas" areas cover 64 percent of the oceans' surface, and are the largest habitat for life on Earth, comprising more than 80 percent of the global biosphere. Despite their vastness, they are significantly impacted by human activities.

Most pressing are the threats from industrial-scale fishing, which has already reduced populations of swordfish, marlin, tuna, sharks, and other large marine predators by more than 90 percent in just 50 years. By-catch of albatross, leatherback and loggerhead sea turtles, and other vulnerable species threaten their extinction within decades. High-tech fish-finding electronics and modernized bottom-fishing gear have opened up new deep-sea fishing grounds and former refuges, threatening species-rich seabed habitats, such as seamounts and cold-water coral reefs, with destruction.

Other threats stem from the persistent noise created by shipping, military activities, and marine scientific research, which is interfering with the ability of sound-

sensitive species to locate and catch food, as well as impeding their communication with others of their kind. Emerging activities, such as commercial energy projects, aquaculture, CO_2 storage, and bioprospecting lack rules to control their impact.

Improved implementation of existing marine agreements is essential to confront these threats. However, to achieve an integrated, ecosystem-based and precautionary approach to the conservation of high seas biological diversity and productivity, we must update and build on these agreements. Obligations under UNCLOS, the Convention on Biological Diversity and the United Nations Fish Stocks Agreement have been put in place to protect marine ecosystems, conserve living marine resources, and preserve rare and fragile marine habitats. They must be applied consistently throughout the high seas. These agreements relate to regulated activities such as shipping, deep-seabed mining, and most conventional fishing activities. They also relate to unregulated activities such as military activities, (many) deepwater fisheries, shark fisheries, and all other human activities that may impact on life and ecosystem processes in our global commons.

Kristina Gjerde is an expert on deep-sea and environmental law and an advisor to the IUCN Global Marine Programme on high seas governance. She received a degree in law from the University of New York and was awarded a Pew Fellowship in 2003 to investigate, highlight, and promote opportunities to improve high seas governance.

Coral reef awash at the water level,
Papua New Guinea

The Coral Reefs

Coral are relentless builders. For 400 million years, generation after generation, inch by inch, these tiny animals have succeeded in a still unequaled achievement: building edifices that can be seen from space. Their secret? Teamwork. A branch of coral gathers together several hundred "persons," polyps that resemble little anemones and share the same calcium skeleton. The growth of the reef is terribly slow, on the order of three feet per thousand years. Attacked by parrotfish or worn down by the waves, the reef is eroding at the same time as it is growing, producing enormous quantities of sand that accumulate on white sand beaches.

The coral reef, along with the tropical forests for which it is the undersea counterpart, is the richest ecosystem on the planet. In these oases of life, hundreds of thousands of species live, bathed by the warm, limpid waters of the tropics, wandering in a multicolored labyrinth. Paradise? Certainly not for the reef animals! Here, parasites proliferate, predators abound, and competition is fierce. Freed from environmental constraints, the animals dedicate all their energy to exploit, devour, evict, and dominate each other. Here more than anywhere else, hell is the others.

Polyps (3/16 inch) of a **stony coral** (3 feet), Mozambique Channel.

Polyps [1/8 inch] of **staghorn coral** (6 feet), Mozambique Channel.

Corals are colonies of thousands of tiny polyps
that share the same calcium skeleton, patiently secreted from the calcium
carbonate dissolved in the water. Each of these polyps is a complete, separate
animal that strongly resembles a small anemone and is composed of a cavity
used for digestion, topped by a mouth surrounded by tentacles.

Polyps (3/8 inch) of **pineapple beaded coral** (6 feet),
Mayotte, Mozambique Channel.

Mushroom-shaped pinnacle at Zélée Bank (20 feet tall), Mozambique Channel.

The atoll-like island of Maupiti, French Polynesia.

Coral reefs appeared in the oceans more than 450 million years ago.
Since then, they have not stopped building gigantic edifices:
breathtaking slopes plunging down into the ocean depths and coral pinnacles
whose shapes can become exuberantly luxurious.
Long after Maupiti Island disappears, the corals will continue
to subsist here, because they will go on growing despite the erosion.

Living Monuments

The "roses of Moorea,"
massifs of **montipora coral**
several feet in diameter,
French Polynesia.

The Night Owls and the Early Birds

Flashlight fish (4 inches), Sudan.

When night falls, the coral reef changes appearance. The herbivorous and omnivorous fish go to sleep, while many of the carnivorous fish wake up to go in search of a meal. Each of these fish species has its own strategic advantages for finding food, or for escaping predators. Spinefoot fish gather together and go to sleep pressed against each other, so that each can transmit an alert message to all the rest in case of attack. As for the flashlight fish, they have a pocket filled with luminescent bacteria under each eye. The light allows them to attract and detect their prey.

Marbled spinefoot and **dusky spinefoot** (8 inches), Jordan.

Long-armed rubble crab (8 inches), Papua New Guinea.

Mushroom coral
(4 inches in diameter),
Mozambique Channel.

The Triumph of the Inseparables

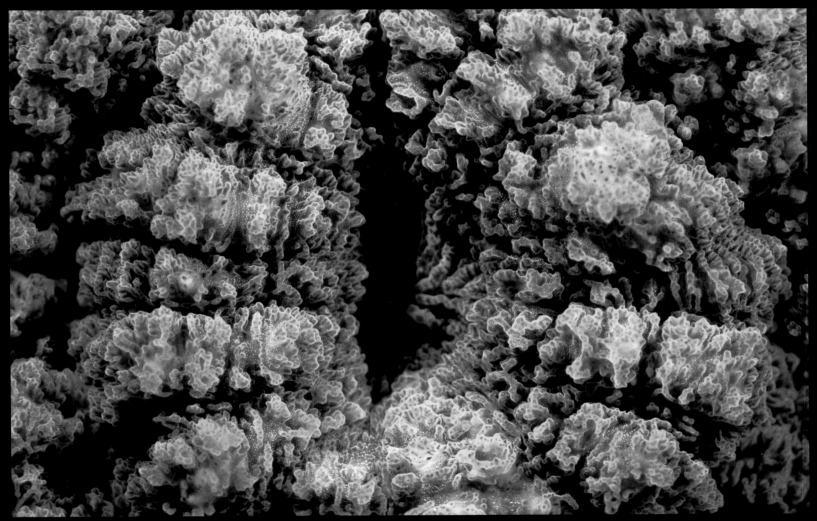

Polyp of red brain coral (2 inches), Mozambique Channel.

There are up to two million single-cell algae in one square centimeter (0.155 square inch)
of coral tissue. These "zooxanthellae," which are smaller than .0004 inches, contribute oxygen and food
to the corals, while in return they absorb the CO_2 and the waste produced by the polyps.
From this perfect symbiosis are born such gigantic edifices as the Great Barrier Reef of Australia,
which stretches over more than 1,500 miles.

Antler coral
(massif 20 inches in diameter),
French Polynesia.

Red-spotted guard crab (1 inch), Papua New Guinea.

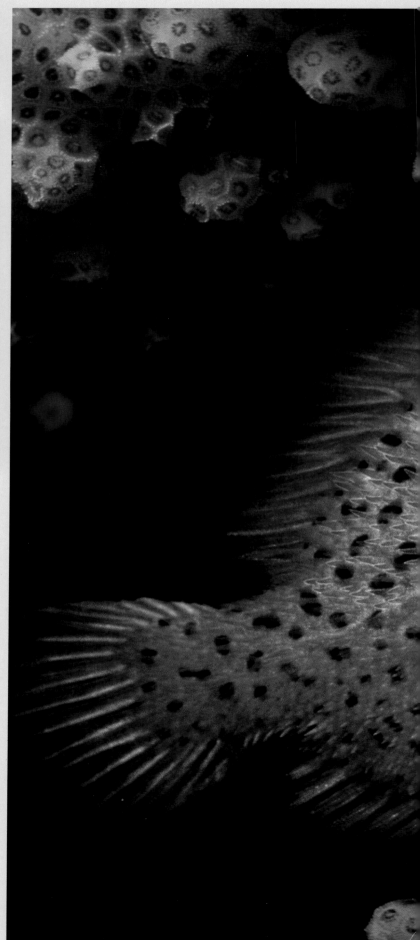

The maze of branches of the antler coral is a microuniverse
that shelters an animal community in miniature.
Spotted coral crouchers, red-spotted guard crabs, and also many
other species, which are practically undetectable, live here
permanently day and night.

Spotted coral croucher
(2 inches), Papua New Guinea.

Male and female sea goldies (6 inches), Mozambique Channel.

The reef that pushes toward the surface and the islet that erodes, Papua New Guinea.

While the island is eroding and its grottoes are being hollowed out,
the coral is developing and extending right up to just below the surface.
In a few million years, there will only be a ring of coral here.
It is a fascinating paradox: Life, which is mortal by definition,
is able to hold fast and endure, while even the most solid of rocks
breaks up and disappears.

The Corals
Defy Time

**The pink algal crest of the reef
of the Fakarava atoll,**
French Polynesia.

Hawksbill turtle (3 feet), French Polynesia.

The spur and groove zone of the coral reef (6 feet wide), French Polynesia.

The Vandals of the Reef

Green humphead parrotfish (4 feet), Malaysia.

Two antagonistic forces are at work on the reef: the accretion of the coral that makes its growth possible, and the erosion that destroys it. On the reef ridge, where the swells from the open ocean break, the waves hollow out deep furrows and make this erosion quite visible. But the continual destruction of the reef is not just due solely to the waves. Many animals attack the coral, too. The hawksbill turtle breaks colonies off to feed on the little sponges pushing beneath the branches. Parrotfish bite directly into the coral to feed. After the coral is digested, it will be expelled in the form of white sand; it will accumulate—and may even contribute to the formation of beaches.

Giant morays (8 feet), French Polynesia.

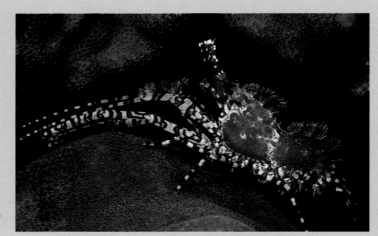

Saron shrimp
(4 inches), Jordan.

Poison goby (2.5 inches), Jordan.

Coral reefs are the habitat of tens of thousands
of species of fish and invertebrates.
During the growth of the reef, shelters of
all sizes are created in the interlaced branches of
the colonies and are immediately inhabited
by a multitude of little animals.
This diversity of dwellings favors biodiversity.
Coral reefs and tropical forests are the two peak
examples of a rich ecosystem.

Yellowspotted scorpionfish
(2 inches), Papua New Guinea.

Coral Reefs: A Precious Asset in Peril

Coral reefs are probably the most charismatic eco-systems on the planet, yet they only cover about 0.09 percent of the world's ocean surface. Admired for their beauty and biodiversity, coral reefs also provide food, construction materials, and shoreline protection for about 500 million people worldwide. Coral reefs are estimated to provide the world with $375 billion in goods and services each year.

A recent transformation

Modern coral reefs started forming about 8,000 years ago after the last ice age. During the peak of the Ice Age, the sea level was 360 to 394 feet below current levels, such that there were only narrow coral reefs fringing the tropical continental shelves. Earlier low sea levels also made it easier for human migrations over land bridges to former islands, where there were unexploited resources. These early arrivals, such as in the Pacific Islands, did not cause major ecological disturbances and the reefs remained relatively intact, with healthy coral, fish, and invertebrate populations until World War II. Many major battles took place over Asian and Pacific reefs, and in the following years, there was also a major growth of human populations in tropical countries.

A capacity for regeneration

Coral reefs have developed under a range of natural stresses, such as tropical cyclones, freshwater floods, earthquakes, volcanic eruptions, tsunamis, and low levels of plagues and diseases. While these could cause considerable local damage, the reefs retained a strong potential for recovery and usually there was little sign of damage 20 years after the event.

Coral reefs under threat

Coral reefs have not been coping well with chronic human pressures over the past 50 years, and there is now a developing crisis around the world. Overexploitation is the greatest threat, especially destructive fishing and mining of corals and sand. Activities on land pose the next greatest threat, with increasing rates of sediment and nutrient pollution caused by growing populations in coastal areas.

If these pressures were not serious enough, there is now an increasing range of global threats driven by global climate change. Increased sea surface temperatures are stressing and killing corals through bleaching. This occurs when corals lose their symbiotic algae, the zooxanthellae that provide them with most of their energy, resistance to UV radiation, and their color. CO_2 concentrations in the seawater are also increasing, making it more acidic and reducing calcification in corals, coralline algae, and mollusks. To make matters worse, it now appears that the increased incidence of coral diseases and plagues of coral predators, such as the crown-of-thorns starfish, are also linked to climate change.

Sadly, coral reefs may be one of the world's first ecosystems to collapse under these pressures. We estimate that about 20 percent of the world's coral reefs have been so badly damaged that they are unlikely to recover, and about 50 percent of all the remaining coral reefs are under such intense pressures that they could be lost in the next 20 to 40 years. The rate of decline is increasing as human populations grow and put growing pressures on reefs. The damage and threats are most severe in Southeast and South Asia, along the coasts of Eastern Africa and in the Caribbean.

What can we do?

There is an urgent need to conserve coral reefs to safeguard their valuable resources. Action is needed to put large areas of reefs aside under full protection, where fishing and damaging activities are prohibited, but non-damaging activities such as sustainable (or environmentally sensitive) tourism are encouraged to provide alternative incomes. The current estimate is that about 30 percent of the reef area should be protected to ensure that there are adequate sources of healthy larvae, and that those areas of reefs chosen for protection are resilient to climate change.

The protection of coral reefs can be advanced by:

- the promotion of non-damaging activities on coral reefs, such as environmental tourism;
- the prevention of dynamite and cyanide fishing for the live fish trade;
- the control of pollution and deforestation in tropical catchment areas near coral reefs;
- the reduction of greenhouse gases that are driving climate change; and
- aid for developing countries, from the governments of rich countries and international agencies, to protect coral reefs.

INDIAN OCEAN: THE IMPORTANCE OF CORAL REEFS

Coral reefs of the Indian Ocean extend along coastlines that support some of the world's densest human populations, including the Indian subcontinent, to some of the least populated oceanic atolls and desert shores of Arabia. They are connected to the Pacific via Indonesian waters. As a result, many species of corals and fish are comon to both the Pacific and the Indian Oceans. However, the Indian Ocean coral reefs have developed under the influence of the monsoon winds and also have many endemic species.

The coral reefs of New Britain (Papua New Guinea) are thriving, but the temperature would need to increase by only a few degrees for the corals to die off massively.

The El Niño phenomenon

Coral reefs in the Indian Ocean are threatened by increasing local and global human pressures. Until 1998, the greatest threats to coral reefs in the region were overfishing and pollution. But the severe El Niño[1] event in 1998 raised seawater and air temperatures, which resulted in the most severe mass coral bleaching event ever recorded. There was a 50-80 percent mortality of corals throughout the Indian Ocean, affecting all reefs, including those in protected areas and remote from human pressure. Scientists from the region now estimate that 37 percent of the Indian Ocean's reefs are severely damaged, and 43 percent are under serious threat in the next 30 years.

Coral reefs: a barrier worth saving

The importance of coral reefs and their vulnerability to threats was illustrated on December 26, 2004, during the Indian Ocean tsunami. The tsunami caused massive damage to coastlines around the Indian Ocean and reached as far as the Seychelles and Africa, killing more than 300,000 people in one day. Coral reefs fringing the coastlines were severely damaged in many places, with 500-year-old coral heads being overturned and vibrant coral communities turned to rubble. Some of the reefs helped protect the land, and now they should be protected to continue to provide for fisheries products, support for the tourism industry, and a barrier to protect shorelines, and people, from the effects of erosion.

David Obura is the East Africa coordinator of the Coral Reef Degradation in the Indian Ocean (CORDIO) program. Based in Mombassa, Kenya, he is specialized in the impacts on coral reefs (fishing, climate change, etc.) and their responses (bleaching).

PACIFIC OCEAN: EMPOWERING LOCAL MANAGEMENT

Coral reefs are spread throughout the Pacific Ocean in Melanesia, Micronesia, and Polynesia in the territorial waters of more than 20 small island states containing thousands of islands with coral reefs and lagoons. These cover 19,000 square miles, with some growing on old volcanoes and others fringing larger islands. People had settled on all of these islands over 1,000 years ago, to such an extent that coral reefs are now an integral part of the culture of approximately two million people.

Reef resources provide them with food and protection from the sea, especially against cyclones. In addition to fishing, they also harvest mother-of-pearl oysters, giant clams, trochus shells, and sea cucumbers as part of their traditional practices. More recently, tourism and pearl oyster aquaculture have emerged as major industries. Although pollution is not high in the Pacific, there has been a considerable increase in sedimentation, overfishing, and urban development of small islands, and many coral reef lagoons are now polluted with sewage.

The reefs of the Pacific, however, are relatively healthy with less than 4 percent severely damaged and with low levels of long-term threats, other than coral bleaching due to global climate change. Pacific reefs will also suffer from increases in human populations and damage due to outbreaks of the predatory crown-of-thorns starfish.

Management of coral reefs under scrutiny

Pacific peoples are now aware of the increasing human threats to coral reefs, and are developing monitoring and management programs. Unfortunately, many governments do not recognize the fragility of their coral reef resources or are diverted by the pressures of development. Others recognize the problems, but lack the necessary human and economic resources to implement comprehensive reef conservation. Therefore, there is a need for a regional approach to such environmental problems, with smaller countries

pooling their resources and gaining assistance from the international community.

Clive Wilkinson is the coordinator of the Global Coral Reef Monitoring Network (GCRMN), the organization that, every two years, publishes a report on the status of the world's coral reefs. As a researcher at the Australian Institute of Marine Science, he regularly collaborates with the IUCN Global Marine Program.

Bernard Salvat is a professor at the Ecole Pratique des Hautes Etudes (EPHE) at the University of Perpignan, France. He is one of the top specialists on Polynesian reefs and manages the Research Centre and Observatory for Island Environment (CRIOBE) in Moorea. He is an active member of the French Committee for IUCN.

[1] An El Niño is a cyclical weather phenomenon characterized by an abnormal rise in the temperature of the ocean.

Dead coral reef (Mayotte, Mozambique Channel): After the El Niño of 1998, which caused a sudden rise in water temperature, more than 70 percent of the corals bleached, then died. It will take at least 30 years (without another episode of increased heat) for the coral to return to its earlier level of health.

English lady crab (4 inches), France.

The Law of the Strongest

The silence of the ocean gives a sensation of peace—but that's only an illusion. Incredible violence reigns here. Right from birth, every creature undertakes a fierce battle to survive in a world of competition where the balance of power and the balance of life are constantly challenged. At every moment, each animal must feed itself while at the same time avoiding becoming the meal. From the sea slug to the shark, from the crab to the manta ray, all animals have evolved by developing formidable weapons for killing, poisoning, weakening, attacking, or defending themselves. Many strategies exist for feeding: whether they filter their food, graze it, or hunt it, each species has its own advantages but also its own weaknesses that its competitors seek to exploit. One single law applies to all: Eat without being eaten, reproduce, survive. Too bad for the weak, the young, the sick, the lost: They're tracked down, crunched, digested. The life of marine animals is not one of leisure or ease. In nature, there is no pity, no generosity—but, on the other hand, there is also no greed or contempt.

The Food Chain

An example of the food chain in the Mediterranean

Mermaid's wineglass (3/4 inch), France.

Plants start off any food chain; these are the only organisms capable of capturing light from the sun.

Elysia (3/4 inch), France.

The algae are then consumed by the herbivores.

European green crabs (2 inches), France

The food chain proceeds, each time with more and more powerful carnivores.

An example of the food chain in a coral environment

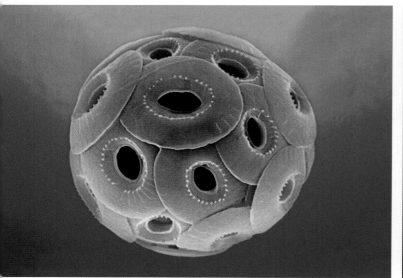

Coccolithophorid phytoplankton (0.00031 inch), North Atlantic.

Colorized image from a photograph taken with an electron microscope. Phytoplankton algae constitute the basic element for many food chains.

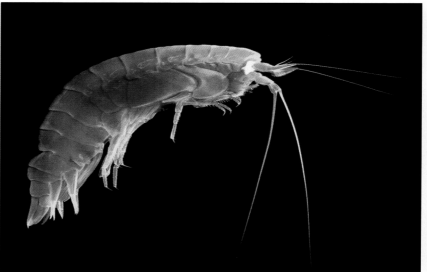

Zooplankton: Unidentified red amphipod (3/8 inch), subantarctic region.

Amphipods and copepods are the most abundant animals on the planet. They feed on phytoplankton.

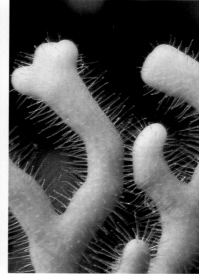

Fire coral (3 feet), Egypt.

With its stinging filaments, fire coral can capture and feed on zooplankton.

Common octopus (3 feet), France.

Dusky grouper (3 feet), Spain.

Porites coral and **melon butterflyfish** (6 inches), French Polynesia.
A narrow mouth allows the butterflyfish to enjoy the coral polyps.

Commerson's frogfish (12 inches), Jordan.

For scientists, a food chain is primarily a flow of energy. Plants capture energy from the sun and transform it into living material usable by herbivores, then by carnivores. At each level of the food chain, only 10 percent of the energy is actually transmitted, because the remaining 90 percent is spent by the animals through their movements or through their reproduction. This is why, in ecosystems, the large predators at the end of the food chain are much less numerous than the herbivores.

Variegated lizardfish (10 inches) and sea goldie (6 inches), Jordan.

A Low-Speed Chase

Articulate harp shell (2.5 inches) and **hermit crab** without its shell (1 inch), Papua New Guinea.

The harp snail sets off in pursuit of a hermit crab, which, sensing danger,
flees and even abandons its shell to gain speed. Alas, this last-ditch effort is in vain.
After a slow but inexorable pursuit, the extendable mantle of the predator covers the
victim, which is immediately devoured.

Lemon shark (11 feet) and **double-saddle butterflyfish** (5.5 inches), French Polynesia.

The Weapons of the Predators

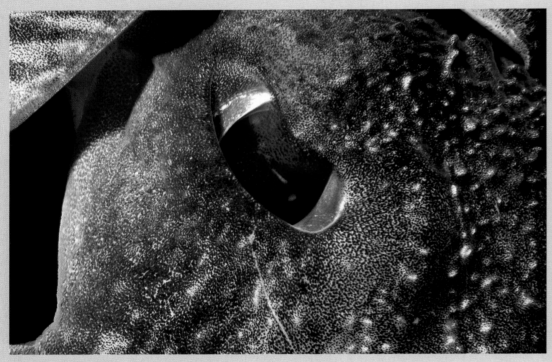

European cuttlefish (16 inches), France.

Predators need to have sophisticated body features to detect their prey, physical strength to launch their attacks, and effective tools with which to seize their victims. So these animals are easy to identify. The cuttlefish's highly developed eyes and the moray's muscular body, huge mouth and sharp teeth reveal that they're both carnivores.

Laced moray (10 feet), Mozambique Channel.

Ouch!

Wolf-eel (6 feet) and **red sea urchin** (10 inches), Canada.

Black-and-yellow rockfish (16 inches), Canada.

Predation is one of the leading causes of death
in marine animals, especially when they are young.
To protect themselves from this scourge, more
than half of fish species have adopted the
strategy of the sea urchin: a body form that makes
them as inedible as possible. The prey of
the coral hind has solid spokes on its fins which
prevent it from being swallowed.

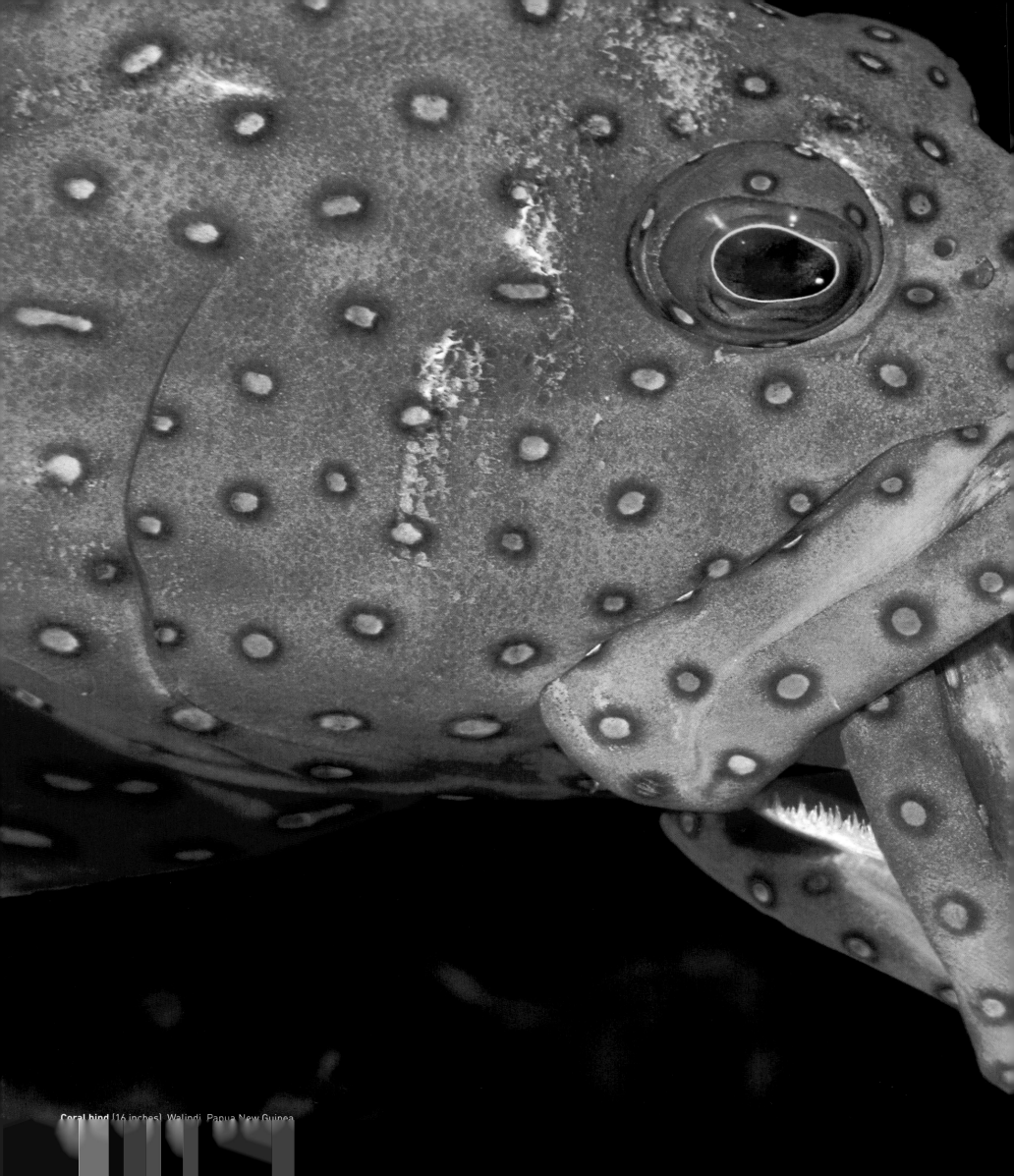

Coral hind [16 inches], Walindi, Papua New Guinea

Defending One's Territory

Coral hind (16 inches), Walindi, Papua New Guinea.

Many male fish jealously guard the territories
they use for feeding or reproduction.
By each one being limited to a precisely defined area,
they reduce the risk of finding
themselves in a conflict that could weaken or kill
them. Occasionally, though, two adversaries may
claim the same territory. If one is stronger than the
other, a simple intimidation maneuver may
be enough to make the weaker one flee.
But if both adversaries are of appreciably the same
strength, a fight may ensue, with each of the two
protagonists being sure of victory.

Clarke's triplefin (3 inches), Australia.

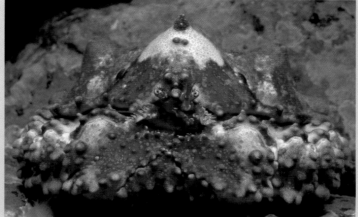

Puget Sound king crab (6 inches), Canada.

With its perfect camouflage, solid shell, and
retractable limbs that fit together to form an
impregnable shell, the Puget Sound king crab appears
to be totally protected from predators. But
unfortunately, as with all crabs, its shell does not grow
with it, and the crab must abandon it by sloughing it
off at regular intervals, leaving the crab destitute
of any defense system. During this molting period, the
crab has only one solution: to burrow into a crevice
and not budge until its new armor becomes solid.

European cuttlefish (16 inches) and **grass goby** (8 inches), France.

A Soft Body but a Strong Mind

Bobtail squid (1 inch), France.

Octopuses, cuttlefish, and squid are without doubt the most advanced invertebrates of the animal kingdom. Scientists have observed that these creatures were able to solve simple problems and remember the solutions they had come up with. With keen eyes worthy of vertebrates, a well-developed nervous system, and an unequaled ability to change color, they are also equipped with an organ that is both powerful and accurate: the sucker, which, in some species, is even armed with venomous barbs.

Suckers of the giant Pacific octopus, Canada.

Yellowbanded sweetlips (16 inches), Papua New Guinea.

Dark-margined flagtail (6 inches), French Polynesia.

Moontail bullseye (10 inches), French Polynesia.

When a creature is weak and at the mercy of more powerful animals, it's better to form a group and face the threat together. A good number of species, such as the moontail bullseye and the dark-margined flagtail, apply this principle to the letter. There are considerable advantages to this strategy: It reduces the risk of an encounter with a predator, makes it possible to spot the predator more quickly because of the many eyes watching, and, if a predator is seen, disrupt its attack by making quick, abrupt movements

European pond turtle (10 inches), France.

Wearing Armor

Some aquatic animals are equipped with solid armor and have no fear of predators.
From the pond turtle to the mitten lobster of the Marquesas,
these animals have to carry their heavy defenses with them wherever they go,
which considerably reduces their mobility. In nature, everything is a matter of compromise.

Marbled mitten lobster (6 inches), French Polynesia.

European lobster
(18 inches), France.

Amphipod (1/2 inch), Canadian Arctic.

Sharks: Essential Predators

For most people, sharks are a reminder of the film classic *Jaws* and viewed as a serious and ever-present threat to individuals daring to venture into their environment. While it is true that a few shark species are potentially dangerous to people and huge publicity attends the relatively small number of shark attacks every year, the true picture is very different. It is the sharks that are in crisis, worldwide, as a result of man's activities.

A fragile ecosystem

Large predators play a vital role in maintaining ecosystems. Removing sharks does not necessarily mean that their prey populations will increase, enabling fishermen to catch more fish; rather, the loss of top predators may cause population declines lower down the food chain. This is be-

Silvertip shark (7 feet), Polynesia. A hundred million sharks are caught each year, or about three per second. The sharks are fished mainly for their fins, which are consumed in soups in Asia.

cause sharks also eat other predators, thus indirectly protecting fish stocks that are valuable to mankind. So, killing sharks may cause an unpredictable range of undesirable side effects for other fisheries and could even damage entire marine ecosystems.

Sharks have been at the top of the marine food web for hundreds of millions of years. They have few natural enemies and need produce only very few young to maintain a stable population. Young may take decades to reach sexual maturity and some sharks live for over 100 years but only give birth every two or three years. The small numbers of offspring of such long-lived, slow-growing animals cannot possibly replace the tens of millions killed in fisheries.

Shark extinction

Major environmental changes in the world's oceans have triggered some massive extinction events during the last 400 million years, but today mankind is responsible for a new extinction event in the world's oceans. This is driven by unsustainable exploitation of marine resources, exacerbated by climate change. Other major threats to sharks include habitat alteration, damage, and loss from coastal development, pollution, and the impacts of fisheries on the seabed and food species.

Significant depletion dates from increased coastal

fishing with artificial fiber nets in the 1950s. Most large-scale target shark fisheries only really began to expand in the mid-1980s as demand increased for their fins (as an ingredient for shark fin soup), meat, liver oil, and cartilage, while by-catch rose as fishing pressure increased generally. These fisheries are largely unmonitored and unmanaged, but huge declines in formerly widespread shark stocks have been identified; some have been depleted by over 90 percent, while population declines of 70 percent over the past 20-30 years are the norm.

Lack of government action

Today, the case for shark conservation is more compelling than ever before. We now understand the importance of ensuring that shark fisheries are sustainably managed in order to yield long-term benefits to coastal communities, whether reliant upon commercial fisheries, sport fishing, or marine ecotourism. The IUCN Shark Specialist Group is working to review the threatened status of all species of sharks, rays, and chimaeras for the IUCN Red List of Threatened Species and to promote the sustainable use, wise management, and conservation of all the chondrichthyan fishes (the sharks, rays and chimaeras) at international, regional and national levels. The urgency of introducing effective shark conservation programs is widely acknowledged by United Nations fisheries and environment agencies, and national governments, but little progress has been made. Lack of action arises from the traditionally low economic value of shark fisheries and the small volumes of products produced, as well as the poor public image of animals viewed primarily as dangerous marine predators. Shark eradication programs have, in the past, been better resourced than shark conservation and management programs!

Sharks, beyond conventional wisdom

Sharks are characterized by a skeleton of light, flexible cartilage. They appeared over 400 million years ago, since evolving into extraordinarily complex and sophisticated animals. All have five to seven paired gill openings on the sides of their head, one or two dorsal fins, and skin protected by sharp scales that constantly fall out and are renewed, just like their teeth. Sharks have no swim bladder, but a large oil-filled liver provides buoyancy. They have large brains, comparable to birds and mammals, and acute senses of smell, taste, vision, and pressure; they are able to detect the tiny electric fields emitted by living animals, inanimate objects, and water moving through the Earth's magnetic field.

Sharks exhibit an extraordinary variety of body forms and colors, feeding and reproductive strategies, and are found from the very deep ocean to intertidal estuaries and large tropical rivers and lakes. Most sharks are about three feet long, but they range in adult size from one foot to 66 feet. Many are fish- or squid-eaters, swallowing prey whole, while others crush and eat shellfish. Some have large triangular teeth for biting chunks out of big prey, including marine mammals, but three of the biggest are plankton-feeders with tiny vestigial teeth. Certain species undertake remarkable long-distance journeys that may

last over a year. Others undertake regular seasonal migrations, but many are very poor swimmers (or use their paired fins to walk on the seabed) and never move far. Many sharks can pump water over their gills to oxygenate them, but the most active species (some of the fastest animals in the sea) force water over their gills as they swim and may die if they stop. Some are pack animals, others primarily solitary.

Shark life history and reproduction is far more varied and often more advanced than that of birds or mammals, although sharks provide no parental care after depositing their young in nursery grounds where larger predators are scarce. Sharks developed internal fertilization, live birth processes, and placental reproduction before mammals evolved. All produce a few large yolky eggs (similar to bird's eggs), which develop into large well-developed pups that can fend for themselves. Pup development follows different strategies. Some sharks lay eggs with tough protective capsules that hatch from as little as a few months later to over a year later, but the majority give birth to large live young after a pregnancy that may last over a year.

Sarah Fowler is a co-chair of the IUCN Shark Specialist Group (SSG), IUCN's global network of shark experts. She is managing director of Nature Bureau International, and a Council Member of English Nature.

Sharks of the Mediterranean

A diverse population of about 82 chondrichthyan species (cartilaginous fish) is present in the Mediterranean. These include four endemic species, 37 Atlantic species, 35 cosmopolitan species, and 2 migrants from the Red Sea via the Suez Canal.

Some species such as the great white shark, the basking shark and the Mediterranean manta ray are already protected under the Barcelona Convention. Others are included in the IUCN Red List, in the appendices of the Bern and Bonn Conventions, or have been proposed for inscription in the CITES[1] appendices.

Current conservation measures are focused on particular species, neglecting some aspects in regard to their role within the ecosystem. Habitat protection and the sustainable management of resources should also be taken into account in the conservation of biodiversity. For this, cooperation is necessary at national, regional and international levels between different jurisdictions, fishers, conservation and environmental bodies, recreational, and game fishing associations, scientific and research organizations.

For the conservation of cartilaginous fish, the following priorities are recommended:

- Full protection status for rare, endemic, and endangered species;
- Development of fisheries management programs to minimize waste and discards and to encourage full use of dead sharks instead of just one body part: the fin;
- Identification and protection of critical habitats, such as nursery grounds, spawning grounds, and mating areas;
- Development of national and Mediterranean research programs, and the raising of public awareness through educational programs and information.

Fabrizio Serena is head of the regional agency for Environmental Protection in Tuscany. He is a shark specialist and collaborates with the IUCN-SSC Shark Specialist Group and the IUCN-WCPA Specialist Group for Mediterranean Marine Protected Areas.

[1] Convention on International Trade in Endangered Species of Wild Flora and Fauna.

India and Indonesia have the largest shark fisheries in the world, but the entire stock of sharks around the globe is under threat. According to a recent scientific study, most of the shark species in the North Atlantic declined by more than 50 percent in less than 15 years.

Adapting to Their Environment

Ninety percent of the history of life on Earth has developed under water. That is where our planet's first living beings appeared some four billion years ago, and that is where they developed over the three and a half billion years that followed. From the first archaic bacteria to the most highly evolved dolphins, marine animals have never ceased to improve and diversify. Despite changing climates and shifting continents, increasingly sophisticated creatures colonized the oceans.

This evolution did not occur smoothly. Major extinction crises occurred, each time resulting in the rapid disappearance of most of the flora and fauna at that time. The most well known is the one that caused the disappearance of the dinosaurs, but other crises had occurred before. It is estimated that 95 percent of the species that existed at one time or another have now disappeared.

The fluid and adaptable last; the fixed and immovable vanish. In a world in perpetual evolution, survival means ceaseless change in order to maintain subtle balances within the body to cope with outward conditions. Anything goes to achieve this result: adaptations of body form, of physiology, or even of lifestyle. Marine animals, like all living beings, are geniuses at performing a balancing act.

Garpike (30 inches), France.

Resisting the Force
of the Waves

For marine animals, the area where the waves hit the shore is quite inhospitable.
However, as in all difficult habitats, competition is not as strong there and access to food is easier,
which is not unimportant. Some sea urchins of the Cape Horn have adapted to resist the swells
and have no spines on the top of their body, in order to improve their hydrodynamism.
Sea snails, for their part, have a different asset: A powerful sucker beneath their heads allows them
to stick tightly to their substrate.

Dufresne's sea urchin
(3 inches), Chile.

Montagu's sea snail
(4 inches), Canadian Arctic.

Fish That Don't Know How to Swim

Grunt sculpin (3 inches), Canada.

Two-stick stingfish (10 inches), Jordan.

Many fish live permanently on the seafloor and have lost much of their swimming abilities
in exchange for developing organs more suitable for walking. The stingfish no longer has a swim bladder,
which is intended to regulate floatability, and its pectoral fins have been transformed into feet.
The grunt sculpin no longer knows how to swim. It is clad in rigid plates that provide it with effective
protection but hinder its movements.

Stonefish covered with seaweed (15 inches), Jordan.

Becoming Part of the Scenery

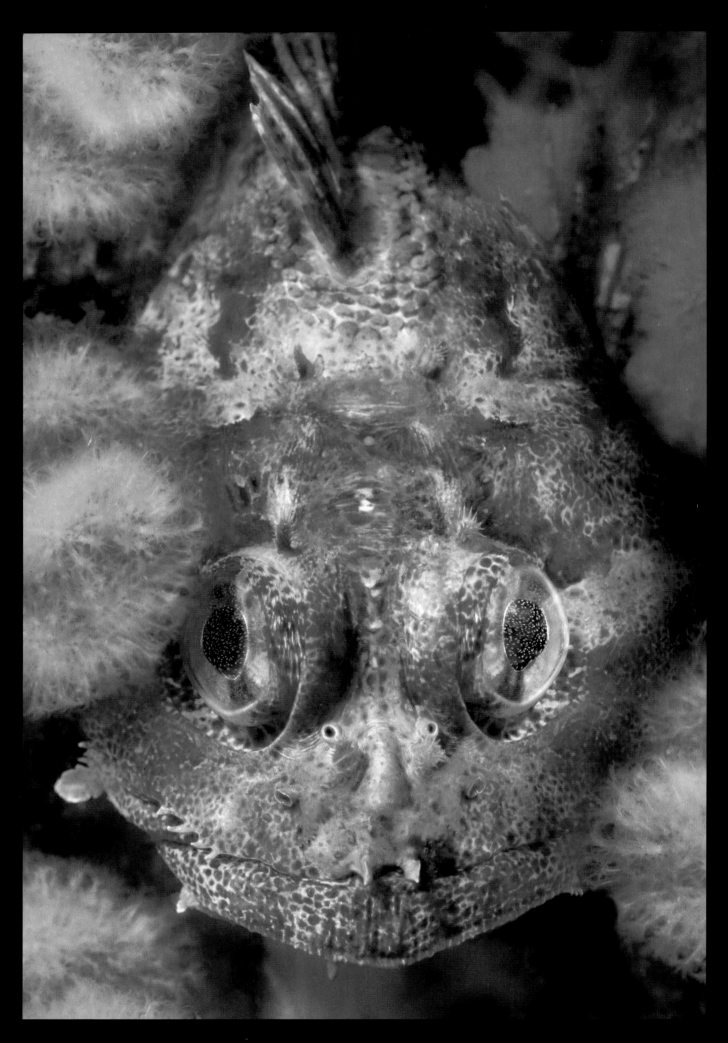

Red Irish lord (20 inches), in sea strawberries, Canada.

To live a happy life, live hidden. That's the motto of the red Irish lords, adept at camouflage, for whom disappearing is a vital matter. Depending on the color of the environment where they live, the color of these fish vary, going from a dominant red to a dominant white. The stonefish's hiding technique is also quite well developed: Its rough skin provides an ideal support for algae that grow there and completely camouflage this redoubtable hunter.

Red Irish lord (20 inches), in giant plumose anemone, Canada.

Postlarva of the emperor angelfish
(1/2 inch), French Polynesia.

Pubescent emperor angelfish
(5 inches), Mozambique Channel.

Some marine animals completely change their way of life between the juvenile stage and adulthood. This is the case with the emperor angelfish, whose larvae are planktonic and grow in the oceanic desert while the adults live in the coral reefs. Escaping from predators, finding a sexual partner, the emperor fish have varying needs depending on their age, and four totally different colors during their lifetime allow them to adapt to each.

The Childhood of the Emperor

Juvenile emperor angelfish
(2 inches), Malaysia.

Adult emperor angelfish
(14 inches), French Polynesia.

Olive sea snake (5 feet), Papua New Guinea.

The olive sea snake is a peaceful animal that travels along the reefy bottoms in search of small prey.
This hunter is well equipped in its quest for food: It has a flexible, hydrodynamic body, permeable skin that allows it to breathe
space out the number of times it needs to go up to the surface for oxygen, but above all it has a powerful venom that makes it o
the most poisonous snakes in the world. Its bite would kill a person in a few minutes, but it has only ever bitten fish.

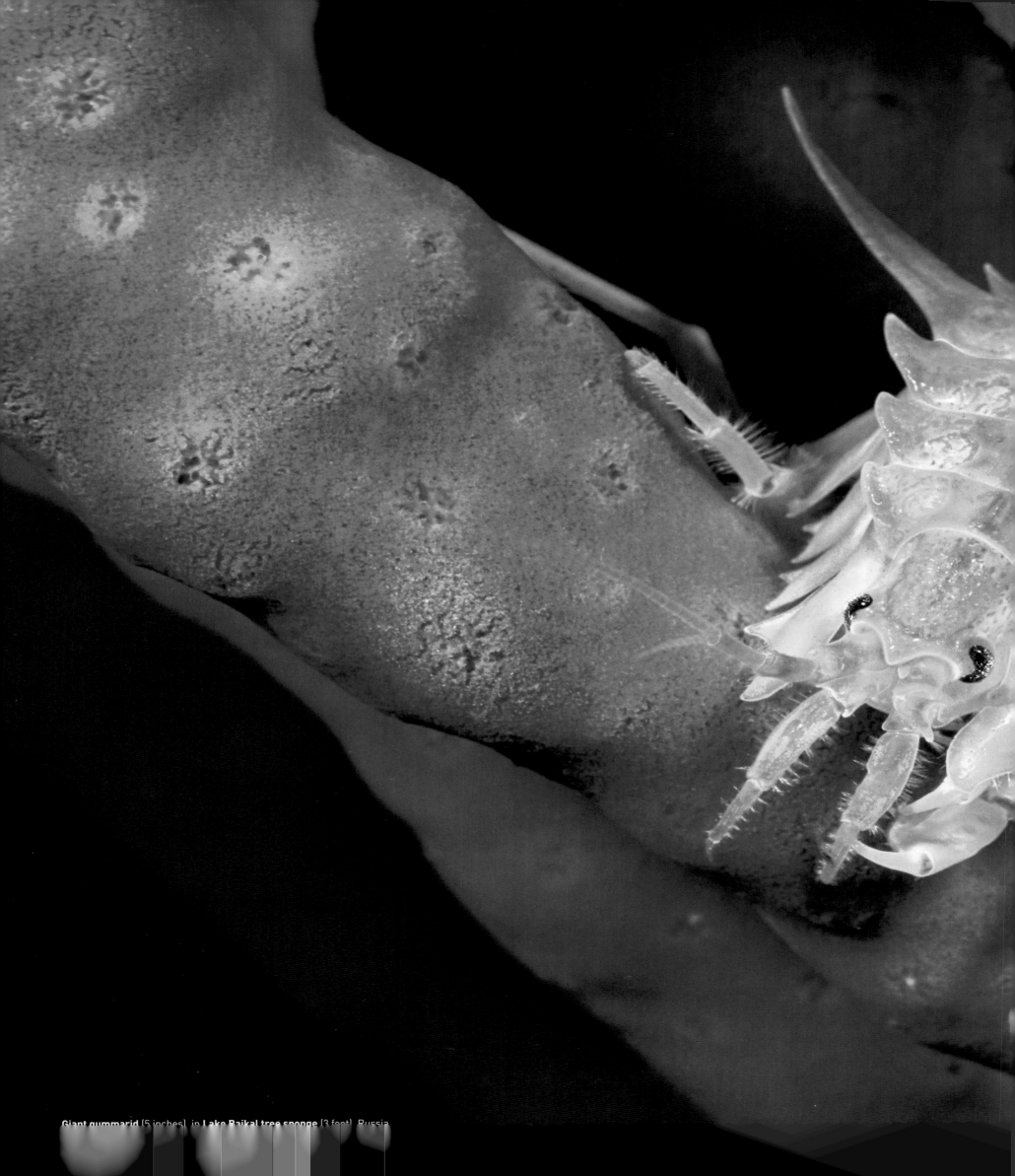

Giant gammarid (5 inches) in Lake Baikal tree sponge (3 feet), Russia

The Invention of the Claw

Skeleton shrimp (2 inches), Canadian Arctic.

The lobster has two very different claws.
One is solid and strong, used for crushing, while the other is sharp and is intended for cutting.
Such sophisticated tools are the fruit of a long evolution.
Originally lobsters were amphipods (literally, "those with differentiated legs"),
such as the arctic skeleton shrimp, which have developed rudimentary claws.

European lobster
(18 inches), France.

Russian Dolls

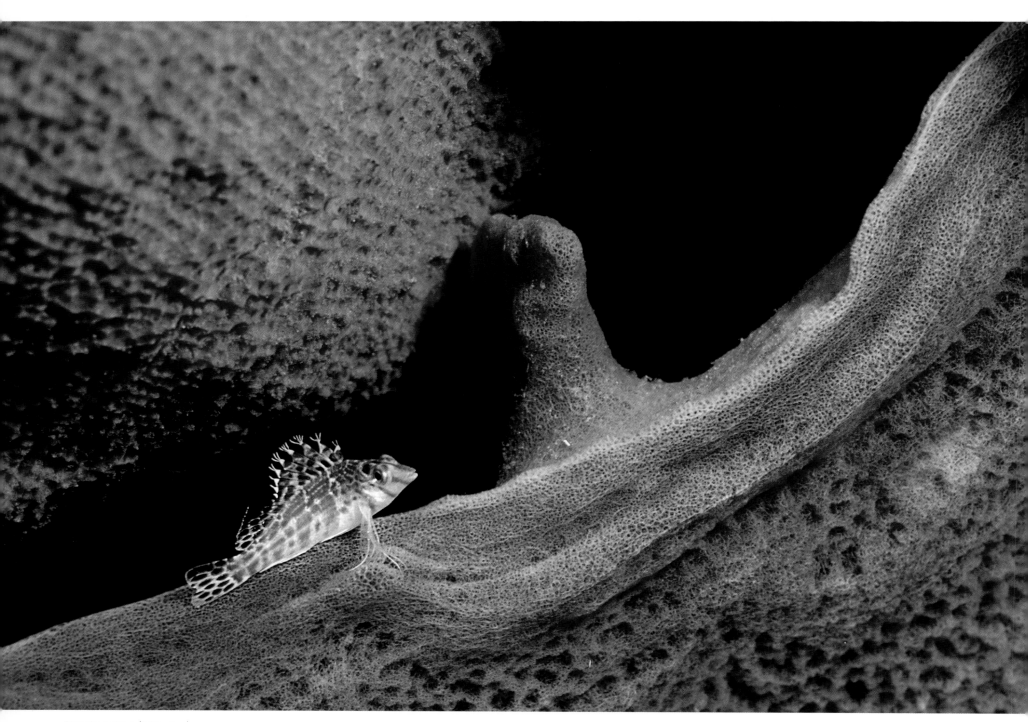

Dwarf hawkfish (2.5 inches)
on a **giant barrel sponge** (3 feet high),
Papua New Guinea.

Giant barrel sponge
(3 feet high), Mozambique Channel.

Biodiversity is multilayered. Just when you think you're observing an organism,
you're actually observing an ecosystem. A simple sponge or coral is also the sole universe of dozens of little animals
that will live their entire lives there, finding refuge, food, and even shelter for their little ones.

Mental wrasse (3 inches) in
arborescent alcyonarian soft coral (30 inches),
Jordan.

Green arborescent alcyonarian soft coral
(30 inches), Jordan.

Paeony bulleyes (10 inches), Egypt.

Red Camouflage

In the first few feet beneath the surface,
warm colors, red then yellow,
are absorbed by the water and everything
becomes uniformly bluish. To blend in with
this monochrome universe, deepwater fish,
such as the Mediterranean wrasse, frequently
take on reddish hues that are bright beneath
the light of headlamps, but are totally colorless
in the absence of artificial lighting.

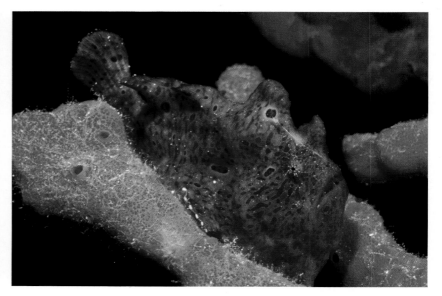

Longnose hawkfish (5 inches) in **alcyonarian soft coral / carnation coral** (3 feet), Mozambique Channel.

Commerson's frogfish (12 inches), in **red sponge** (3 feet), Jordan.

Wrasse (6 inches), in **red gorgonian** (3 feet), France.

The Marine Iguana—
Survival Hero

San Bartolomé Island, Galápagos Archipelago, Ecuador.

Don't wish an iguana's life even on your worst enemy.
Behind a placid appearance, the marine iguana owes its survival
only to an extraordinary capacity for endurance.
In order to reach the seaweed that is the essential part of their food,
the large males must dive and can lose up to 50°F in body heat while they are
immersed. The females and their young,
unable to withstand such cold, must content themselves
with grazing on the meager carpet of seaweed that
appears at low tide. It gets worse: During times of food shortage,
90 percent of the iguana's food disappears, and the iguanas are decimated.
To stay alive, the survivors then begin to shrink, including their skeleton,
reaching sizes that can be as low as 20 percent of their initial weight.

Marine iguana (3 feet), Ecuador.

Red sea cucumber (8 inches), Canada.

A Meal for the Red Sea Cucumber

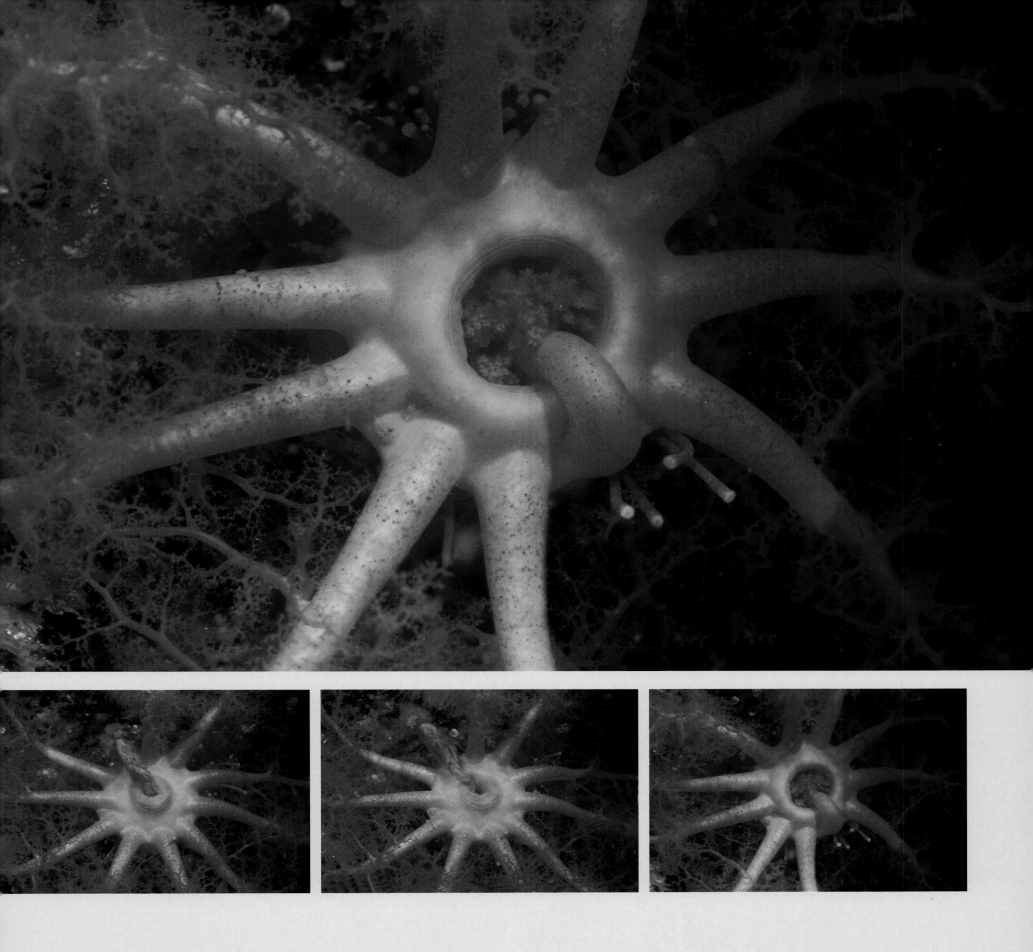

The red sea cucumber feeds on myriads of tiny organisms, such as
larvae and eggs that float in the seawater. Its technique is highly perfected:
It deploys its long, finely branched arms in the current, then brings them back
one by one to its mouth to gobble the fruits of their harvest.

Giant gammarid (15 inches), in **Lake Baikal sponge** (3 feet) Lake Baikal, Russia.

Baikal, an Evolving Ocean

Cracks in the transparent ice field of Lake Baikal, Russia.
Freshwater gammarid in the ice floe (3/4 inch), Russia.

Lake Baikal today is like the Atlantic Ocean was 150 million
years ago. In a distant future, it will resemble the Red Sea.
Located at the juncture of two tectonic plates that are constantly moving
away from each other, it has been increasing in size for
23 million years. During all that time, the fauna of the lake has remained isolated.
Giant freshwater gammarid ten times larger than their oceanic counterparts,
previously unknown sponges resembling green trees, Baikal yellowfin
pre-dominating in all of the lake's waters, the fauna of the Baikal has
something unique that gives one the feeling each time of
being plunged into another planet.

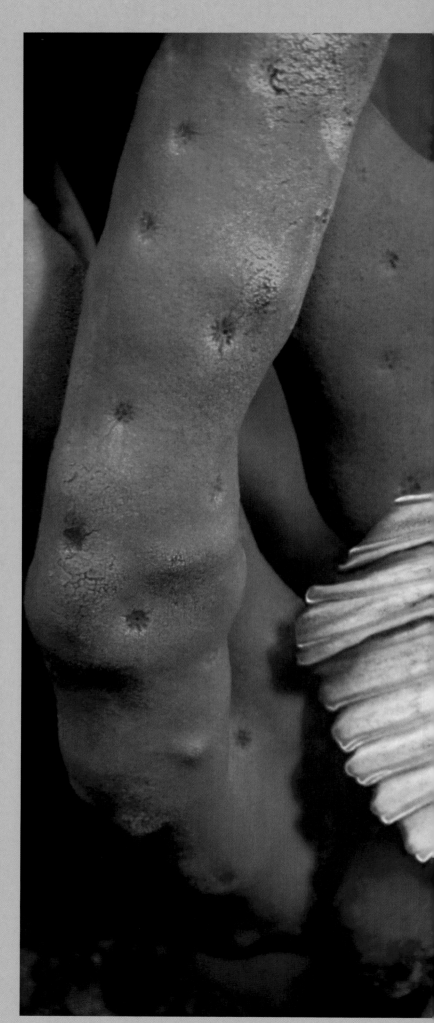

Baikal yellowfin (5 inches), Russia.

Tourism and Marine Biodiversity

The marine and coastal environment offers an infinite number of opportunities for tourism. The sea not only provides many benefits in terms of food and transport, but people have always been fascinated by the beauty of its beaches. It attracts the curious and the adventurous, as well as those seeking tranquility.

Tourism is the largest industry in the world (one in every 13 jobs). Growing faster than other branches, marine tourism is particularly dynamic. As an example, two million visitors each year go to the Great Barrier Reef, which makes tourism one of the greatest contributors to the Australian economy (with sales figures of more than a billion Australian dollars per year).

The explosion of tourism

In the United States, travel and tourism contribute more than $800 million to the gross national product each year, and it is estimated that the coastal states attract at least half of the tourists who visit the country. In spite of the appeal of the famous national parks such as Yellowstone and the Grand Canyon, more than 80 percent of tourism revenues come from the coastal states. A visit to the Florida Keys is enough to appreciate the number of tourists attracted by the subtropical waters.

Tourism has also become a source of support for many of the world's island nations. In the Seychelles, the number of visitors exceeds that of the local population, and the revenues generated increase the standard of living of the inhabitants well above that of nonaligned countries that have no access to the sea.

The strangling of the goose that laid the golden egg

For several decades now, world tourism has undergone an increase that can be attributed to an improvement in the standard of living and the increase in leisure time in developed countries, as well as to lower transportation costs and to improved communications. Unfortunately, though, this rapid expansion has not been without damage. In simple terms, by destroying the quality of the resource it is exploiting, maritime tourism has become an enormous machine under which the goose that laid the golden eggs, is gradually being strangled.

In the past, most of the coasts were protected from tourism by their inaccessibility, safety problems, and the relatively high cost of transport and leisure activities at sea. Today the coasts are subjected to multiple pressures whose cumulative and collective effects can be devastating. Poor planning, lack of control, lowered water quality, disorganized usage of marine resources, erosion of beaches—these are factors that can have a terrible impact on coastal tourism.

In conclusion

Obviously many things can be done. The managers and planners know this well. However, it must be made known by all possible means that the coastal areas are highly integrated and exceptionally fragile ecosystems. The large hotels can claim that their customers are not doing anything but sitting on the beach—a benign activity. But these same customers eat the reef's fish, which are caught by the impoverished local population without any resource management and are sold for practically nothing, thus undermining the possibilities of a sustainable use of the resources. Who will benefit from this? Where is the equitable access to the resources? Who owns the goose that lays the golden eggs? Who steals the eggs before running away?

François Odendaal is based at the Cape in South Africa. He is a specialist in marine protected areas, tourism planning, and the use of coastal resources. He works with the IUCN on management of protected areas, as well as public communication and information.

Tourism is the world's largest industry, providing one in every 13 jobs.

Protected Tourism in the Sinai Peninsula of Egypt

Although tourism is often presented as a possible solution to the conservation of biodiversity and to the challenges of community development, experiences that generate tangible benefits for the local population and the environment are rare indeed.

An ambitious conservation of natural resources

By improving the protection of the environment in order to support the development of tourism, Egypt has offered jobs to the population at the national as well as the local level. The current value of the coral reefs for tourism is estimated at around U.S. $500 per 10 square feet per year. This value is used for evaluating the damage caused to the environment and the fines related to that damage.

In 1989, responding to requests from the Egyptian government, the European Union agreed to provide technical assistance to the Egyptian Ministry of the Environment to develop and manage the first marine-protected area in Ras Mohammed. This project was based on the principle that the coral reefs, of remarkable quality and international importance, should receive specific treatment and legal protection to ensure their survival. They thus were offering a guarantee for the government's economic development programs in the south of the Sinai. In this way, the first Egyptian national park was created, inaugurating what were to become guidelines for an ambitious conservation of natural resources.

Marine-protected areas

From there, Egypt decided to establish a network of land and marine protected areas in order to preserve critical natural resources and support the national economic development policies. The Declaration of Networked Protected Areas in the Gulf of Aqaba has made it possible to establish a large Marine Protected Area (MPA) covering the entire Egyptian coast of the Gulf of Aqaba. The objectives of the government, supported by the commission of the European Union, were achieved: The coral reefs and associated ecosystems in the Gulf of Aqaba are completely protected, measures to prevent any discharge are firmly established, any alteration of the coast is prohibited, local fishing is regulated, and a consensus about the management problems with the resident communities has been reached. The protected areas development program of the south of the Sinai owes its success to strict legislation, constant support from the government and to the establishment of functional partnerships with the local communities.

A model expansion

From 37 square miles of protected land and sea in 1989, there are now more than 4,250 square miles, including five land-protected areas (Ras Mohammed, Nabq, Abu Galum, Taba and Sainte-Catherine) and 200 miles of shores and adjacent land.

The protected areas program was implemented during a period of rapid expansion of the tourist industry in the south of the Sinai. Available facilities in Sharm el-Sheikh have increased from 1,000 beds in 1988 to more than 30,000 in 2002. The maximum development of Sharm el-Sheikh was set at the end of 2002, even though other allocated zones are being developed. The objective fixed for the tourist development of the subregion of the Gulf of Aqaba was 160,000 beds by the year 2002, but the tourist boom ran into commercial and political difficulties.

With the government's financial commitment and the support of the European Union, a new project is being prepared to assist the regional authorities in managing the principal identified threats to the environment (waste water and household garbage treatment).

Alain Jeudy de Grissac, a marine environment expert for more than 30 years, especially for the protected areas and their management in the Mediterranean, West Africa, and the Indian Ocean, is currently working in Eritrea. His work on tourism in the Sinai was carried out in collaboration with Omar Hassan and Michael Pearson.

The impact of diving tourism on undersea flora and fauna is a potential problem that should not be ignored.

Pair of peacock blennies (8 inches), France.

The Love Life of Marine Animals

The first marine organisms (bacteria) that appeared four billion years ago used a technique for reproducing that was simple, effective, and very boring. They divided into two equal parts. This technique is successful (it is still in use today), but it has two major drawbacks: it does not permit the development of complex organisms; and it produces only perfectly identical clones that die off en masse at the slightest change in the environment.

A billion and a half years later, some groups of animals made a fantastic breakthrough: the egg, the product of sexual reproduction between two "parents." By mixing the genes of two partners, the egg makes it possible to multiply infinitely the diversity of individuals, which caused the stunning acceleration of natural selection processes. By grouping together in an immature "body" all of the genetic information needed for its development, the egg results in the development of complex organisms.

The love life of marine animals ultimately has only one purpose: to provide a future for the egg so that they can disseminate their genes. From the selection of their partner to the raising of their young, they dedicate an enormous amount of energy to performing this task. The appearance of the egg is perhaps the most striking event in the history of life on Earth.

Cutting bud of the brooding anemone (4 inches), Canada.

Brooding anemone (4 inches), Canada.

Astonishing Variety

Mating of sea hares
(6 inches), France.

Mated pair of bluecheek butterflyfish (9 inches), Egypt.

Reproduction of banded dye-murex (2 inches), France.

Marine organisms, perhaps more than their land counterparts, have developed a great variety of techniques for reproducing.
From the simple asexual cloning of the anemone to the faithful mated pairing of a butterflyfish, and everything in between:
Transitory couples formed by sea hares, gigantic coral beds serving as singles clubs for eagle rays, and mass
gatherings of snails during mating season.

Spotted eagle rays (5 feet), French Polynesia.

The Secret of the Spotted Ratfish

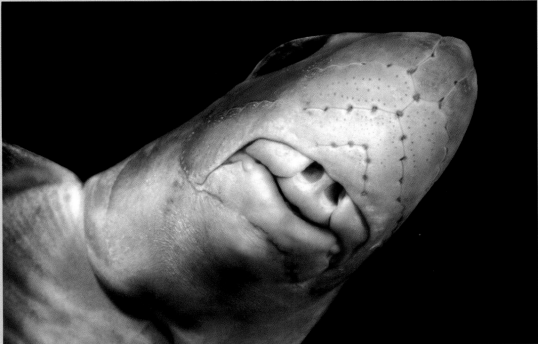

Spotted ratfish (3 feet) and **giant plumose anemones** (1.5 feet), Canada.
Mouth of the spotted ratfish, Canada.

Spotted ratfish are very strange, very ancient fish, whose skeleton is cartilaginous like that of sharks.
They have a venomous spine on the top of their body, and their mouth is armed not with teeth, but with grinding
tooth plates. On their forehead, the males have a white hook called a tenaculum, covered with denticles. Scientists long
wondered about the utility of this organ. The mystery is now solved: It is used to grasp the female during mating.

Spotted ratfish
(3 feet), Canada.

Eyes in the Eggs

Cabezon (3 feet), Canada.

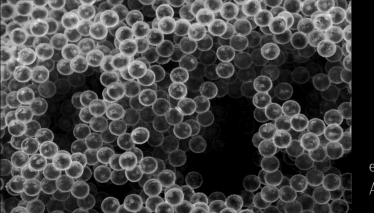

Male cabezons jealously watch over their offspring. Woe to intruders who would get too close to their eggs—they will be mercilessly attacked. Several clutches of eggs may be laid side by side, and the maturation of the eggs is then clearly visible. At first they are a homogenous red, then turn lighter and lighter. After several days of incubation the eyes of the alevins become visible.

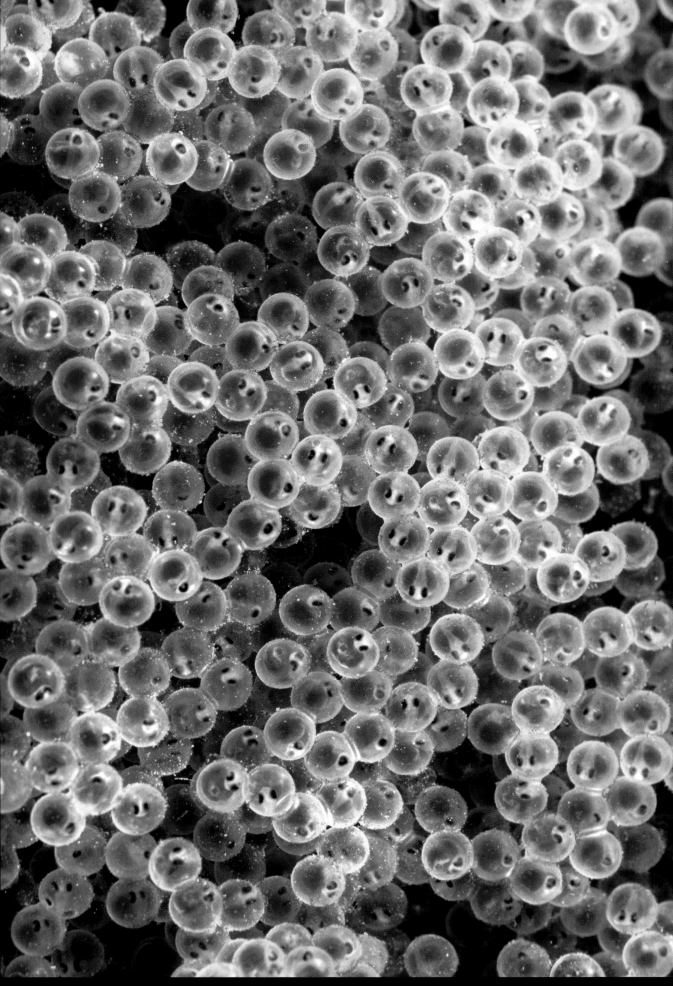

Mature eggs of the cabezon (0.15 inch), Canada.

Gray reef sharks (6 feet), French Polynesia.

The Teeth of Love

Gray reef sharks coupling (6 feet), French Polynesia.

Tiputa Pass, French Polynesia.

For 15 years he dreamed of it, but Yann Hubert would have to make nearly 10,000 dives in order to become the first person to film a mating of gray sharks in its entirety. His pictures have made it possible to understand the biology of this species better. In an immobile position during mating, the sharks suffocate and only owe their survival to the passing current that ventilates their gills. The females are particularly afflicted because they are beset by a succession of males that want to mate with them and bite them to get ahold of them and keep them in position for the mating.

Naturalist guide Yann Hubert, French Polynesia.

Common seadragon with its eggs (16 inches), Australia.

The coast of Kangaroo Island, Australia.

With seadragons, it is the males that carry the eggs until they are hatched. When mating season arrives, the skin of a male seadragon's belly softens and becomes spongy. The eggs laid by the female are released along a sticky filament and immediately stick to the father's belly. The male's skin then quickly becomes rigid, protecting each egg in an individual chamberlet.

Papa Hen

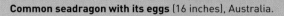

Eggs of the common seadragon embedded in its abdomen (0.2 inch), Australia.

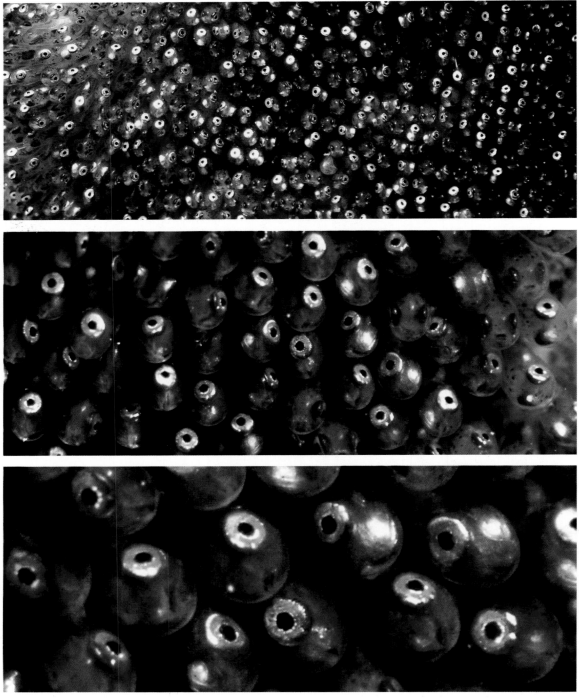

Eggs of clown anemonefish (0.04 inch), Jordan.

A Nursery for Clown Babies

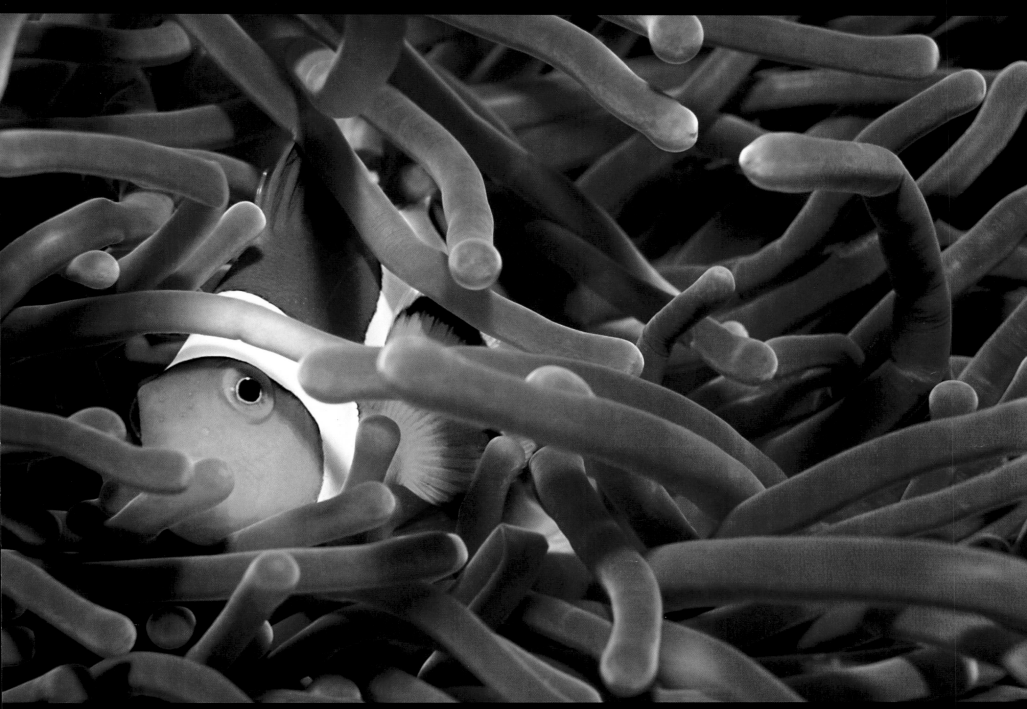

Clown anemonefish (3 inches) in **Ritteri anemone** (3 feet), Thailand.

While they are still only embryos sheltered inside their egg, "clown babies" can already recognize the odor of the anemones that house the members of their species. Thanks to this odor "imprint," the larvae will know how to find the same species of anemones in order to settle down and grow after a planktonic phase spent drifting in the ocean.

The Octopus Is Discreet

Common octopus (3 feet), Spain.

For the octopus, mating is a delicate matter. These extremely territorial animals do not suffer gladly the proximity of a fellow octopus, and the males suffer from a cruel uncertainty: They are incapable of discerning the sex of their partner before starting to mate! However, during mating season the reproductive instinct takes over and allows an octopus to approach another one. But not too close—the male positions himself at a respectable distance, then discreetly inserts his copulation tentacle, which is supplied with a sperm sac, into the body of his partner. It is not until that fateful moment that he will find out if he is dealing with a female or another male. If he has made a mistake, he runs a high risk of a ruckus!

Postlarva of the longhorn cowfish
(0.2 inch), French Polynesia.

Juvenile longhorn cowfish
(1.5 inches), French Polynesia.

The life of the longhorn cowfish begins with quite a hard ordeal: They have barely been laid and fertilized on the coral reef, when the cowfish eggs leave to drift with the tidal currents toward the immensity of the open ocean. After several weeks of a high-risk journey, the few survivors return to the reef and change quickly. They must rapidly equip themselves with sophisticated means of defense, such as bony plates, horns, and spines, in order to have any chance of survival.

Postlarva of the longhorn cowfish seven days later
(1/2 inch), French Polynesia.

Adult longhorn cowfish
(7 inches), Papua New Guinea.

Changing Gender

Female Asian sheepshead wrasse (20 inches), Mexico.

Male Asian sheepshead wrasse (3 feet), Mexico.

Some fish change sex over the course of their lives. While this phenomenon is rare among other vertebrates, it affects about 10 percent of fish families. The order in which they are one sex or the other depends on the species. In the Asian sheepshead wrasse, the individuals start as females, then become males, totally changing in appearance. In other species, the situation is still more complex. Two different types of males can coexist, depending on whether they have passed through a female phase or were born directly as males.

Mating display of the male Asian sheepshead wrasse
(3 feet), Mexico.

Humpback whale (52 feet) and her **newborn**, French Polynesia.

The Burp of the Whale Calf

Dilated throat of humpback whale (20 feet), French Polynesia.

At birth, the whale calf is less than 10 feet long. Born in a gentle tropical climate, it benefits from ideal conditions for growing and preparing to follow its mother to the feeding zones in the harsh polar waters. The whale calf nurses for about a year. During that time, it gains some 150 pounds per day, while its mother loses 10 tons during suckling. The milk, which is extremely fatty, is very nourishing and dilutes poorly in seawater. The whale calf's burp is not a quiet one.

Ribbon of mother's milk
(5 feet), burped up by the whale calf,
French Polynesia.

Cetaceans: Such Fragile Giants

Cetaceans inhabit (or did inhabit until we eliminated or excluded them) all marine waters throughout the world, as well as several large rivers and associated freshwater systems. Their distribution is limited at the poles only by the presence of continuous ice cover. Continents, ice massifs, and more subtle features, such as depth and temperature gradients, current boundaries, and variations in productivity, create barriers separating species and populations.

The modern global cetacean fauna consists of 14 recognized species of baleen whales (mysticetes), in four families, and more than 70 species of toothed cetaceans (odontocetes), including the dolphins and porpoises, in ten families. Some baleen species, such as the blue, fin, and humpback whales, are cosmopolitan. Others, such as the bowhead whale in the Arctic and the pygmy right whale in the Southern Ocean, are regional endemics. Odontocetes, too, include species with extensive worldwide distributions (sperm and killer whales), but also many endemics with highly restricted distributions. For example, Hector's dolphin lives only in the coastal waters of New Zealand, the vaquita is confined to Mexico's upper Gulf of California, and the baiji is endemic to the Yangtze River of China.

Cetaceans and ecosystems

As highly mobile and adaptive mammals, cetaceans exploit a diversity of habitats. The roles they play in ecosystems, however, are not yet well understood. The role of cetaceans in nutrient cycling, maintaining commensal and parasite faunas, and providing food for scavengers has been noted but little explored. One of the more interesting examples of physical structuring of habitat involves gay whales. They disturb the sea bottom (benthos) while feeding on dense beds of amphipods. This could help maintain the sand substrate by suspending fine particles, resulting in higher densities of prey over large spatial scales.

The slaughter of whales

Many cetacean populations have been reduced in both abundance and range, usually because more animals were killed by whalers or fishermen than the populations could replace. Whaling was a holocaust for the world's large whales, beginning with early Basque adventurers in the Middle Ages and ending with the super-efficient factory ships and fast catcher boats that combed the planet's oceans, including the Antarctic, during the 20th century. Only remnants of the world's populations of great whales remained when large-scale industrial whaling ended in the 1980s.

In the past, killer whales, white whales (belugas) and Black Sea dolphins were the objects of culling campaigns intended to reduce perceived competition with fisheries. Nowadays, some politicians, fishing lobbyists, and scientists, particularly in Norway and Japan, advocate renewed or expanded hunts for minke, Bryde's, sei, and sperm whales, citing these species' consumption of harvestable fish and squid resources and the need for humans to orchestrate a "balanced" ecosystem.

Trade or control of predators?

The debates are muddled by many factors besides the sheer complexity of food web interactions. Apologists for the whaling industry deliberately mislead the uninitiated by circulating photographs of whale stomachs full of fish and quoting high rates of depredation on longlines without explaining why these may be unrepresentative. On the other side, many conservationists express concern that fisheries are affecting the food supplies of cetacean populations, but fail to acknowledge that this argument for ecological competition cuts both ways.

Fisheries' traps

Commercial whaling (now often packaged as predator control or as a response to the supposed need for scientific data) still attracts global attention and foments controversy, but the immediate threats to cetacean populations are much more diverse. Incidental mortality from entanglement in fishing gear is the primary threat to numerous species, driving some to the brink of extinction. River dolphin populations are increasingly fragmented by hydroelectric, irrigation, and flood-control projects; river dolphins in some areas are at risk of becoming trapped in canals and other artificial water bodies with no hope of returning to their natural habitat. Some whale populations that were severely reduced by whaling (e.g., North Atlantic and North Pacific right whales) are now vulnerable to low-level mortality from accidental ship strikes or, as in the case of western Pacific gray whales, to a suite of threats from offshore oil and gas development.

Pollution: a planetary plight

Underlying those relatively direct threats are three that are less obvious but perhaps more sinister: toxic contaminants, underwater noise and climate change. In the absence of experimental evidence, scientists and conservationists have been forced to extrapolate from experience with other taxa to infer potential effects of anthropogenic chemicals on cetacean health. It is abundantly clear, though, that because of their high position in the food chain and tendency to retain large amounts of lipid in tissues (e.g., blubber), cetaceans acquire worrisome body burdens of organochlorines and other potentially toxic compounds.

Cetacean conservation at stake

It was recently discovered that beaked whales—a poorly known but diverse group of deep divers—mass strand after exposure to certain types of high-intensity sonar (including some equipment used by the military), and this has made noise pollution a central conservation concern. Finally, the implications of climate change

Drifting gill nets can measure dozens of miles. The nets trap animals indiscriminately, capturing an unacceptable number of dolphins, birds and marine reptiles.

complicate virtually all management strategies for conservation. In coming decades, the conservation of cetaceans will continue to require the best efforts of non-governmental organizations, a strong cadre of cetological experts and effective institutions at local, regional and international scales.

Randall Reeves, a consultant based in Hudson, Quebec, has undertaken numerous research missions on cetaceans worldwide, from Alaska to the Asian rivers. He is also an expert in the history of whaling, and is the Chair of IUCN's Cetacean Specialists Group.

THE WHALES THAT CAME IN THROUGH THE GIBRALTAR FALLS

Six million years ago, the connection between the Mediterranean and the world's oceans was severed when tectonic forces and a drop in sea level combined to cause the Strait of Gibraltar to dry up. Given that evaporation greatly exceeded precipitation in the region, the Mediterranean rapidly became a salty desert. Some 1.5 million years later, Atlantic waters once again spilled into the Mediterranean, due to a relaxation of tectonic forces and a rise in sea level. The Strait of Gibraltar became the site of a gigantic waterfall. This engendered the flourishing of modern Mediterranean fauna and flora, and led Atlantic whales and dolphins to settle in the inland sea.

The cetaceans of the Mediterranean

Today, nine of the world's 85 cetacean species can be found in the Mediterranean. They range from the gigantic 66-foot-long fin whale, found mostly in the western area, to the diminutive harbor porpoise, confined to the northern Aegean Sea. Recent genetic studies confirmed that cetacean populations of the Mediterranean reside there for life, although limited exchanges do occur with populations from the adjacent Atlantic Ocean, the Black Sea, and the Red Sea.

Cetaceans' survival at issue

Human presence at sea has intensified over the last century, causing serious concern for the survival of Mediterranean whales and dolphins; this is despite an absence of organized and deliberate killing of cetaceans in the region. Fishing activities, however, cause unsustainable levels of accidental killing (by-catch), and most probably depletes prey resources for many populations. Intense maritime traffic across critical habitats, such as in the Li-

Although commercial whale hunting has been prohibited since 1982, Japan, Norway, and Iceland continue to have powerful whaling industries, each year catching more than 1,200 large cetaceans despite the moratorium established by the International Whaling Commission.

gurian Sea and the Strait of Messina, is a source of disturbance and mortality through collisions. Powerful military sonar lead beaked whales to panic and strand, while pollution and coastal development result in habitat degradation. As a consequence, common dolphins have been listed as "endangered" by the IUCN Red List of Threatened Species, and several other species may follow suit in the near future.

Finding solutions for coexistence

However, there is still hope. The region's riparian nations have resolved to act on behalf of their whales and dolphins, and a specific regional agreement is currently being implemented with this purpose: the United Nations Agreement on the Conservation of Cetaceans of the Black Sea, Mediterranean Sea and Contiguous Atlantic Areas (ACCOBAMS). The challenge to conserve this particular region's cetaceans is an overriding objective, because proving that whales and dolphins can coexist with humans in such a crowded sea demonstrates that coexistence is possible everywhere else on the planet.

Giuseppe Notarbartolo di Sciara (www.disciara.net) is a marine ecologist. Once president of the Italian Institute of Marine Research, he is now the head of the Milan-based Thethys Institute, a not-for-profit scientific organism specialized in cetaceans. He is also the deputy chair of the IUCN Cetacean Specialist Group and the coordinator of the IUCN World Commission on Protected Areas Mediterranean Group.

Female spinecheek anemonefish (3 inches) in bubbletip anemone (20 inches), Papua New Guinea.

Living Together

Marine animals have a sense of priorities: Eat without being eaten and reproduce as quickly as possible. This is already plenty enough to do to fill up a life.

These basic relationships, however, are far from representing the fascinating variety of interactions between animals. Some of these interactions are mutually beneficial and allow the two partners to become stronger together. Mutual aid, however, has its limits, and in many other cases of animal interaction, one animal exploits another.

At the heart of ecosystems, no species, no animal lives in isolation, and these relationships are often more essential than the living beings themselves. Without their symbiotic algae, coral are incapable of building reefs. Without symbiosis, there would be no lagoons or coral atolls.

The importance of these relationships requires the exchange of easily decoded messages. The shimmering colors of tropical fish are simply signals essential to the cohesion of the group, and are intended as visual clues for their partners and competitors, and for intruders. Could deciphering the relationships that unite marine animals shed some light on us? The time has come to explore marine colleagues and negotiators, as well as marine profiteers, squatters, and exploiters.

Traveling Companions

Striated hermit crab (4 inches) with its **parasitic anemone** (5.5 inches), Spain.

The anemone that the striated hermit crab carries on its back protects the crab with its urticant tentacles. In exchange, the anemone gathers up the remains of the hermit crab's meals. Even though these two partners can live separately, their association is mutually beneficial. This pair, however, seems poorly matched: The hermit crab has undoubtedly aimed too high and taken on more than it can carry—and now can barely move under the weight of this millstone.

Striated hermit crab
(4 inches), Spain.

Common remora (30 inches) end **manta ray** (13 feet), French Polynesia.

Silver pearlfish (4 inches) in front of an **ocellated sea cucumber** (16 inches), French Polynesia.

"A mouse may help a lion," so goes the fable, but it does not say whether manta rays, condemned to put up with the incessant, inquisitive, even impudent presence of remoras, have offered their opinion on the matter. As for sea cucumbers, the upper end of their well-oxygenated intestine commonly houses several "cotenants." Some of these serpent-shaped fish have parasitic tendencies and feed directly from the ovaries or testicles of their host.

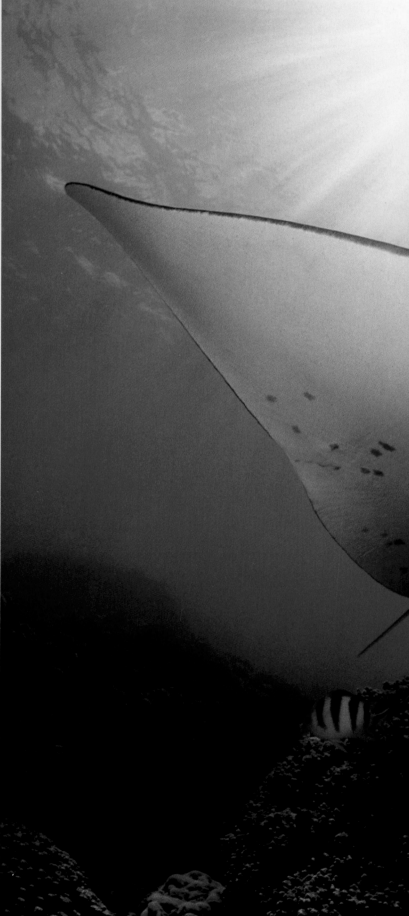

Manta ray (13 feet), French Polynesia.

The Dentists

Bluestreak cleaner wrasse (3 inches) and **midnight snapper** (22 inches), Papua New Guinea.

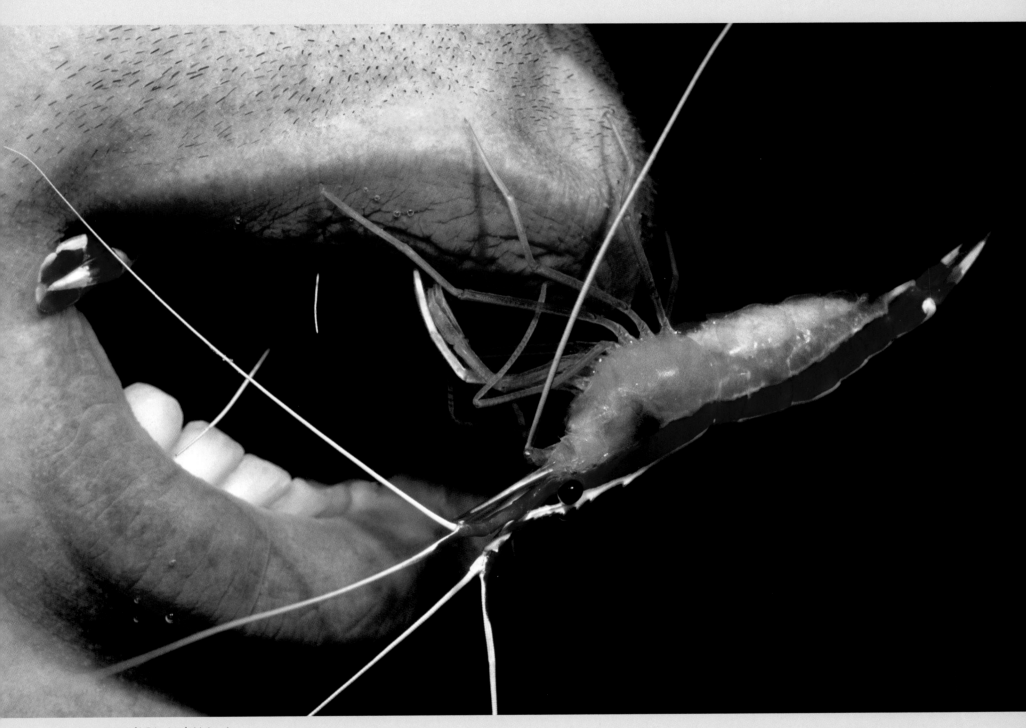

Cleaner shrimp (1.5 inches), Malaysia.

Every animal in the world needs to take care of its personal hygiene, and marine animals are no exception. Since they cannot get rid of their parasites, dead skin, and other oral problems by themselves, they use the services of small species that are specialized in doing such things. The bluestreak cleaner wrasse and the cleaner shrimp can even organize veritable "service stations," where all the large animals, humans included, can benefit from their care.

The Whale Calf Has Lice!

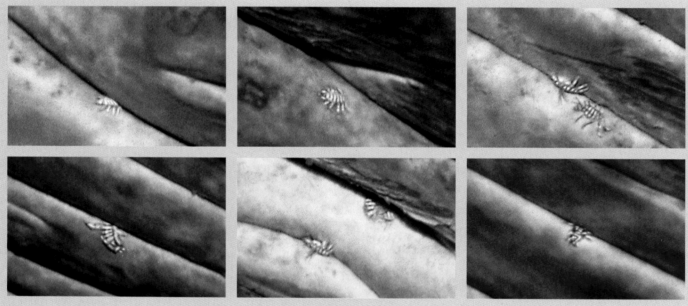

Whale lice (1/2 inch) in the wrinkles of the **humpback whale calf** (20 feet), French Polynesia.

In the folds of the whale calf's skin live large numbers of lice, firmly clamped on to the peaceful giant, thanks to their hooked legs. To get rid of them, the whales can either avail themselves of the services of the cleaner fish, or they can make spectacular leaps that allow them to shake off their parasites and dead skin when their 30 tons of weight slam violently back against the surface of the water.

Eye of humpback whale calf
(20 feet), French Polynesia.

Toothy goby (1/2 inch) on sun coral (5 feet), Mozambique Channel.

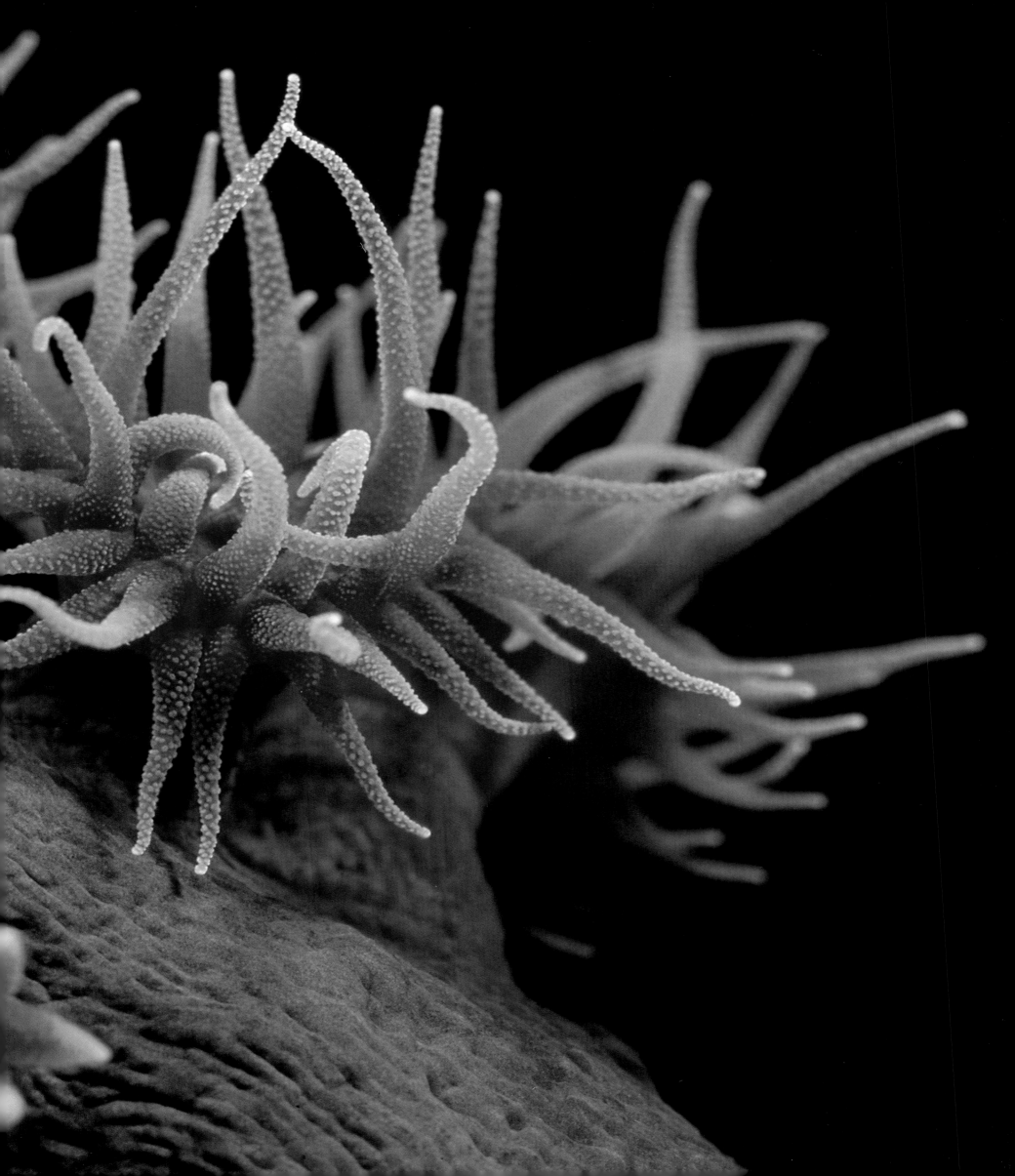

The Squatters

Elegant squat lobster (3/4 inch), Papua New Guinea.

Living at the expense of others is the law of nature for all animals, including humans. Only plants
that exploit the light from the sun and a few obscure inhabitants of the ocean depths are exceptions to this rule.
Some animals, however, seem to go further than others in this concept of exploitation.
The anemone shrimp and the elegant squat lobster camp out with impunity in anemones
and crinoids in order to benefit from the shelter and cover these creatures provide,
gaining these benefits at no cost or effort for themselves.

Anemone shrimp (1 inch)
in **fat anemone** (4 inches), France.

Unidentified cowry (3/4 inch) in alcyonarian soft coral (3 feet), Jordan.

Parasitic isopod (3/4 inch) on the **golden goby** (3 inches), France.

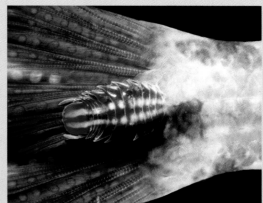

Parasitic isopod (1/2 inch) on the tail of the
East Atlantic peacock wrasse, Spain.

Parasitic isopod on the **striped anthias** (5 inches), Jordan.

Mediterranean sea lice are formidable parasitic crustaceans that suck blood from the fish to which they attach themselves. Their adaptation to life as parasites is especially refined: Two individuals attached to the same fish can communicate and even reproduce by using the circulatory blood system of their host.

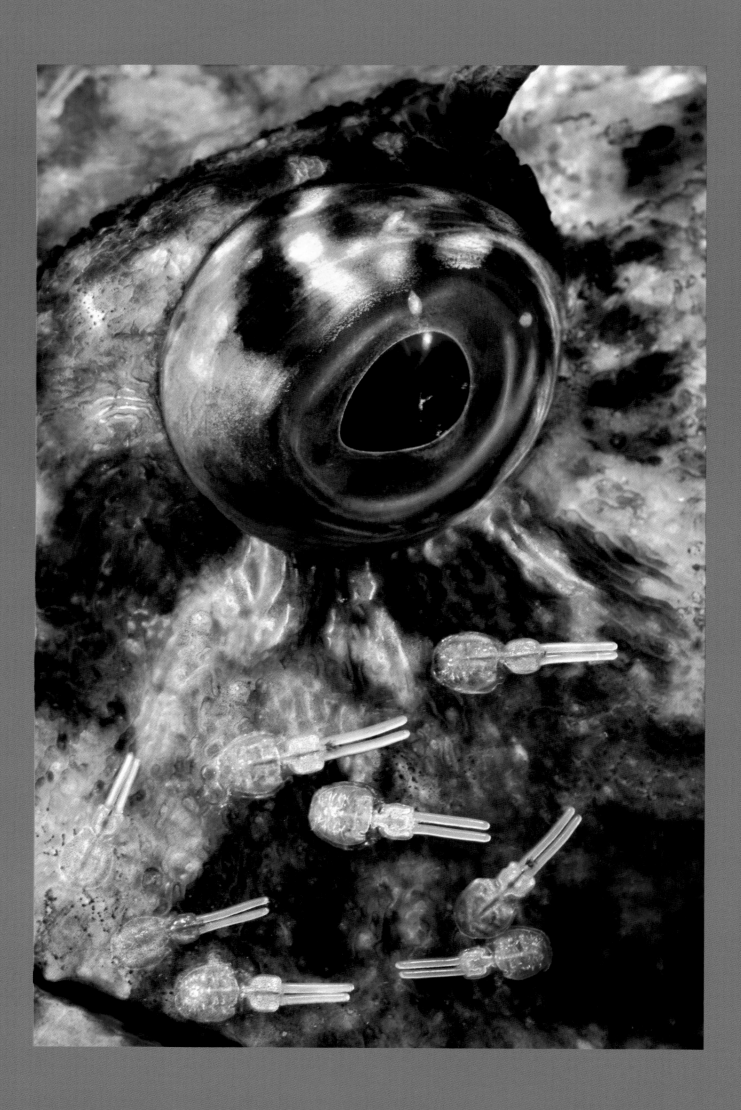

The Parasites

Even though they are often invisible, parasites are an essential component of ecosystems.
Every large wild animal is host to several hundred of them. Generally, parasites do not
seek to kill their host because that would be contrary to their own survival.
Scientists even think that some host-parasite relationships are the source
of symbiotic behaviors that provide mutual benefit.

Cabezon (3 feet), Canada.

Navigation and Biodiversity

Growth in world trade has led to a massive increase in maritime traffic, and the tonnage of goods shipped is also destined to increase dramatically. From an ecological point of view, this is a good thing as this mode of transport is more energy efficient (50 times more than road transport) and requires less infrastructure development. However, it is the source of a variety of pollutants.

Biological pollution

Ballast water is loaded by ships to maintain stability after unloading cargo. It then remains in the ship on its next voyage to pick up new cargo and is discharged in coastal or port waters at its new destination. The ambient ballast water holds living organisms and sediment from the port of origin. Upon discharge in another part of the world, the organisms enter into an ecosystem where they, as nonnative species, do not belong. Some of the nonnative species survive and can establish a permanent population, often at the expense of the host ecosystem. Each year, 106 billion cubic feet of ballast water is discharged in coastal waters and ports. As global shipping is rapidly increasing, and expected to triple in the next twenty years, curbing the associated problems is urgent. Since 1990, ballast water has been on the agenda of the International Maritime Organization (IMO, the United Nations agency for shipping); in 2004, a convention was adopted to prevent the transfer of alien organisms through ballast water from ships.

At present, the transfer of organisms is reduced by exchanging ballast water in open ocean waters. In the future, ballast water will be treated to prevent transfer of organisms. The convention clearly states that the management of ballast water should not cause greater harm than it prevents. Six developing countries, participating in the GloBallast Programme, gained experience in ballast water management, which they now share within their region. GloBallast countries also exchange their knowledge with those in the North.

Organisms can be also carried across the oceans via a ships' hulls ("fouling"). Antifouling paint, containing highly toxic tributyltin (TBT, which is to be banned by IMO Convention starting in 2008), and less toxic alternative antifouling products, can never fully prevent this transport route for organisms. Fortunately, during transport, fouling organisms are exposed to changes in salinity or temperature, which often prevent their survival.

Chemical pollution

Maritime shipping can cause operational or cargo-related pollution. A few decades ago, the world was shocked by the large oil spill that washed up on the French Atlantic coast, following the running aground of a crude oil carrier, the *Torrey Canyon*. This disaster, together with increased concern about the effects of operational oil discharges, creating tar balls along

the equator and oil-polluted beaches, led to the Marine Pollution Convention (MARPOL 73/78) which regulates most ship-related pollution. Apart from crude oil, huge amounts of liquid and solid chemicals, vegetable oils, ore, and container-packed goods are shipped around the world; accidental spills and loss of materials can be equally as harmful as crude oil spills. IMO developed a code of practice for cargoes of liquid noxious substances, well aware that, once a disaster with a chemical tanker occurs, up to 400,000 tons of noxious substances may be released. Shipping disasters occur more often in busy shipping routes, which are often in the vicinity of coastal areas. The pollution associated with a shipping disaster can take a heavy toll on coastal ecosystems, simply due to the enormous quantities of oil or other hazardous materials involved. Frequent accidents off the western European coast, from northern Norway to the Bay of Cadiz underpinned the decision to declare the northeastern Atlantic of Europe a Particularly Sensitive Sea Area (PSSA) in relation to shipping.

When ships become too old, they have to be decommissioned and dismantled. At present, the majority of ships sail to South Asia, where they are beached and taken apart by the local workforce, unprotected against the many hazardous and dangerous materials on board. The hazardous waste runs off freely into the local ecosystem. IMO has recognized the problem and, in spite of great resistance to a change of practice, a few participants in the shipping industry have taken action to develop clean-ship recycling practices that can set an example for the industry to follow.

Noise pollution

In the turbid and dark submarine environment, animals use acoustic and chemical communication rather than their sense of sight. Ships' engines are noisy and the low-frequency sound they generate travels far in seawater. This is often of a similar frequency to the acoustic communication used by marine animals, and tends to be louder, too. No surprise then that numerous cetaceans (whales, dolphins, and porpoises) become disoriented by underwater noise and end up stranded on beaches.

Cato ten Hallers-Tjabbes *is a specialist in maritime issues. She works at the University of Groningen, Holland, and the University of Washington in the United States. She has advised the Dutch government on environmental issues since the 1970s and represents IUCN in IMO Convention meetings.*

Colliding with ships is a frequent cause of mortality among whales. It occurs mainly on the routes of "high-speed" ships, which can reach speeds of up to 50 knots.

INVASIVE ALIEN SPECIES

Invasive alien species are one of the four greatest threats to the world's oceans, along with pollution, the destruction of coastal habitats, and climate change.

Clandestine transport

Organisms can spread rapidly in aquatic environments once they have been introduced. For example, microscopic Japanese algae have recently been found in the North Sea while giant, three-foot-long Pacific crabs are now roaming off the Norwegian coast. Some alien species become invasive to the detriment of local species and can damage the environment or threaten economic activity. An example is the North American comb jelly *Mnemiopis leydii*, which was most probably introduced into the Black Sea and the Sea of Azov with ballast water in the early 1980s. The jellyfish preyed on large amounts of zooplankton, including the young of plankton-eating fish such as anchovies, and by 1994 the anchovy fishery had almost disappeared.

Pathways for the introduction of alien species are numerous. They can be intentional (aquaculture, aquarium-trade purposes) or unintentional (ballast water, hull fouling). The Suez Canal is by far the major pathway for arrival of alien species. The recently announced plans to increase the depth and width of this canal could result in an even greater ecological upset unless some type of salinity barrier is installed.

The risks with deliberate introductions

The deliberate introduction of marine species for cultivation has often brought positive benefits to the economies of many coastal communities, but unfortunately there is a downside. Some of these introduced species have established themselves in the wild and displaced native marine life. Others have brought diseases and parasites, compromising not only native biodiversity but also ecosystem health. To minimize impacts, deliberate introductions should be the subject of rigorous risk assessments.

Marine Protected Areas (MPAs) are not immune to these threats. A boundary on paper will not stop an invasive species. The designation of a site as an MPA may even increase the risk of invasion, as more visitors are attracted to it, and the vessels they travel on can bring invasives species with them. Sadly, even though the World Parks Congress recognized the problem in September 2003, few MPAs are addressing alien invasive species within their management plans.

In the marine environment it is very hard to eradicate an alien species that has become established. Early detection that leads to rapid response is vital. Preventing or fighting invasive species should be addressed at the international and regional level, as well as at the level of national and local concern. International instruments include the International Convention for the Control and Management of Ships' Ballast Water and Sediments, adopted in February 2004, and the ICES (International Council for the Exploration of the Sea) code of practice on the introduction and transfer of marine organisms.

Of course not all alien species become invasive, but the problem is that it is hard to predict which ones will cause problems. Therefore it is necessary to treat all alien species with the utmost precaution.

Maj de Poorter works in the Centre for Invasive Species Research at the University of Auckland, New Zealand. She also heads the IUCN Invasive Species Specialist Group.

Observed for the first time in 1984 off the coast of Monaco, the seaweed *Caulerpa taxifolia* had spread by the end of 2000 to about 25,000 acres in the Mediterranean, impoverishing the sea's ecosystems and obstructing fishermen by clogging their nets. The seaweed is continuing to grow.

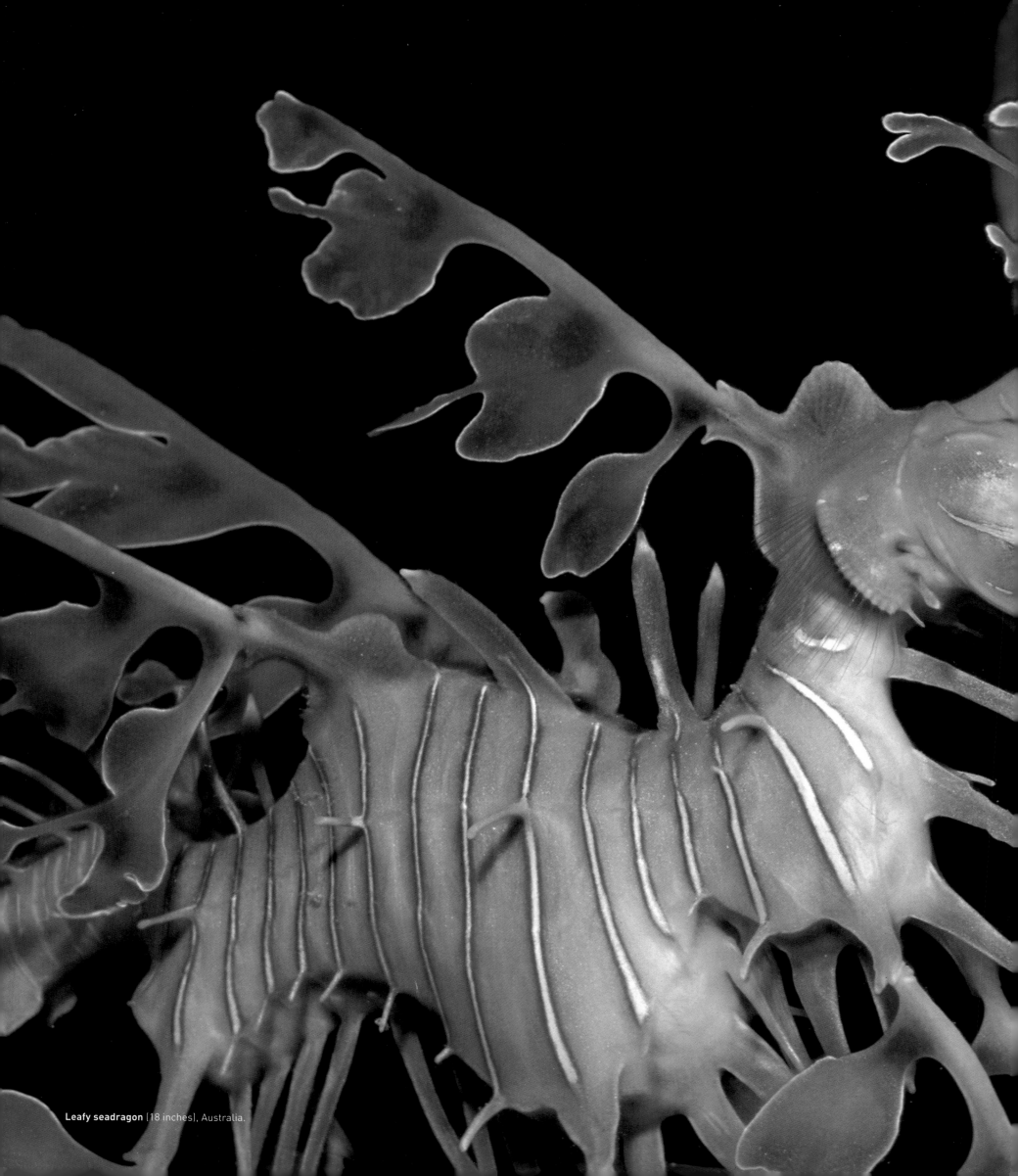

Leafy seadragon (18 inches), Australia.

The Indispensable Oceans

If the marine world fascinates us, it is primarily because it eludes us. No one understands it, no one masters it. It awakens in each of us powerful feelings of freedom, amazement, soothing reassurance, and sometimes also feelings of anxiety or fright.

Perhaps the most troubling aspect is to realize that this "other world," so far from us, holds within it our past, and helps determine our future. At this beginning of a new millennium, modern man seems to have rediscovered the forgotten links of yesteryear that tied us to Mother Nature. There are countless instances of these links, and science is constantly adding to them, expanding our understanding of the process involved in climate regulation, oxygen production, and the water cycle. Scientists remind us each day that we depend on the equilibrium of great natural forces, and that this equilibrium is threatened by our industrial society. Here is perhaps the greatest paradox of ecology: How to imagine that we can have an impact on the planet when "the forces of nature" exceed us so completely?

It took 2,000 years, from ancient Greece to the Enlightenment, for it to be accepted by everyone that the Earth revolves around the Sun. Lets' hope that it won't take us as long to recognize that the future of life on Earth is our responsibility.

Dinoflagellate (0.0015 inch), France. Colorized image from a photograph taken by electron microscope.

Veritable pulmonary alveoli of our planet, the microscopic algae that make up phytoplankton
are responsible for 80 percent of the oxygen production on Earth.

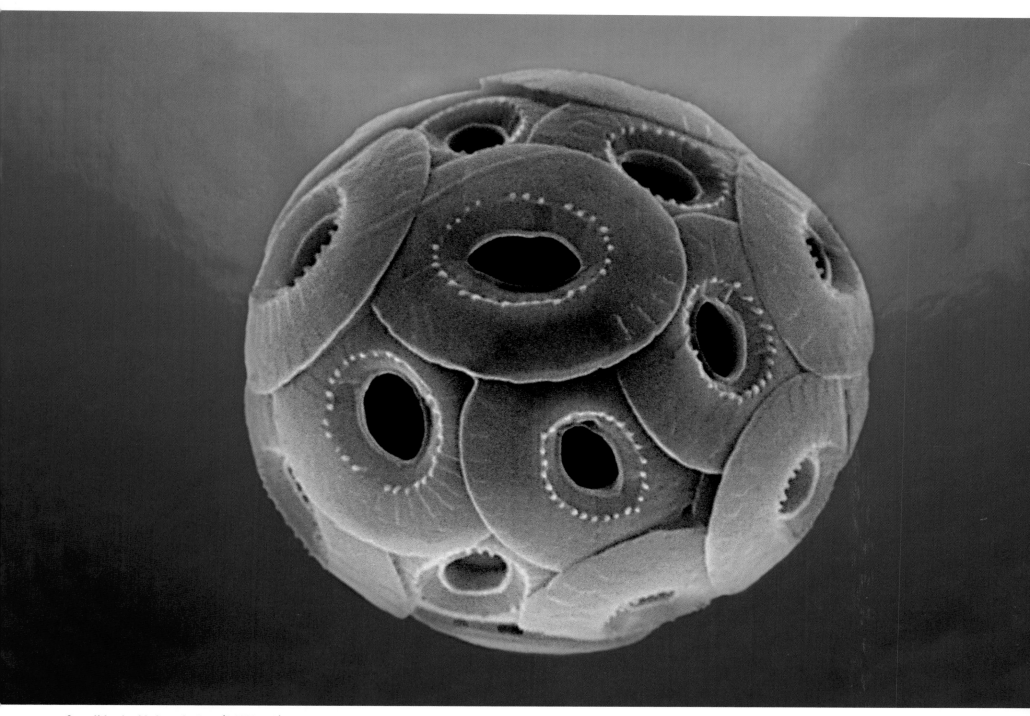

Coccolithophorid phytoplankton (0.0003 inch), North Atlantic. Colorized image from a photograph taken by electron microscope.

Products from the sea comprise the principal source of protein for a billion human beings, and currently represent 10 percent to 20 pecent of animal protein consumed on the planet. Each year, 90 million tons of fish are taken from the oceans, but this quantity has stagnated in spite of constantly improving technological means. What will future generations eat tomorrow if we exhaust the oceans today?

Round sardinella
(14 inches), France.

A Source of Proteins

Spider crab (4 inches) in the **oyster beds**, France.

Oyster beds in the **Thau Basin**, Gulf of Lyon, France.

Farmed red drum (3 feet), Mayotte, Mozambique Channel.

Lightbulb seasquirts (1 inch) and **white seasquirts** (6 inches), France.

Organisms that are fixed in one place, from giant tropical trees
to little seasquirts, are totally incapable of fleeing in case of danger.
To compensate for this handicap, they have developed an extraordinary
diversity of chemical substances in order to repulse parasites, competitors, and predators. Some of these molecules are active against viruses, bacteria, or cancer cells,
and may lead to the preparation of new medicines for mankind.
Currently only 1 percent of marine species have been studied for their
chemical composition. Are the medicines of tomorrow waiting for
us at the bottom of the oceans?

Football seasquirt (16 inches in diameter), France.

Medicines for Tomorrow

**The slope of the
coral reef**, Egypt.

European fan worms (14 inches), France.

Pouting (14 inches), France.

Scientific monitoring of artificial reefs, Gulf of Aigues-Mortes, France.

Sometimes human beings can help nature.
Built on sandy bottoms, artificial reefs re-create a rocky environment
that is richer than the sandy bottoms. The immersion of artificial reefs in the Gulf
of Lyon in the Mediterranean has provided spectacular results:
within a few months, the concrete was completely colonized, and in just three
years the weight of the animals per 10 square feet was multiplied by a thousand!

Common toad (6 inches), France.

From the fountainhead to the sea, a whole series of biological communities that succeed one another
as the streams turn into rivers. When these rivers flow into the sea, they carry nutrients
and sediments that play an essential role in the balance of the coastal ecosystems.
Unfortunately, their impact is not always positive:
75 percent of marine chemical pollution comes from rivers.

Font Estramar spring, France.

Hérault Valley, France.

Common carp (3 feet), released after being caught, Montpellier.

Eurasian minnows (4 inches) in the spring of Buège, France.

1 - **Four-color nudibranch** (1.5 inches), Jordan.
2 - **Meckel's sea slug** (4 inches), France.
3 - **Farrah's sea slug** (1 inch), France.
4 - **Faceline sea slug** (1.5 inches), France.

Baikal green sculpin (unidentified species) (12 inches), Lake Baikal, Russia.

Electric eye scallop (1 inch), France.

A Reservoir of Biodiversity

Giant freshwater gammarid (5 inches), in **Lake Baikal sponge** (3 feet), Lake Baikal, Russia.

Biscuit sea star (2 inches), Australia.

One and a half million species have been inventoried to date on Earth. Among them, only one-fourth are marine creatures, while the oceans offer 250 times more living space for animals than the continents. In fact, this figure shows our profound lack of knowledge of the marine world. Some scientists believe that 20 million species could currently be living in the oceans.

Pigmy leatherjacket (4 inches), Australia.

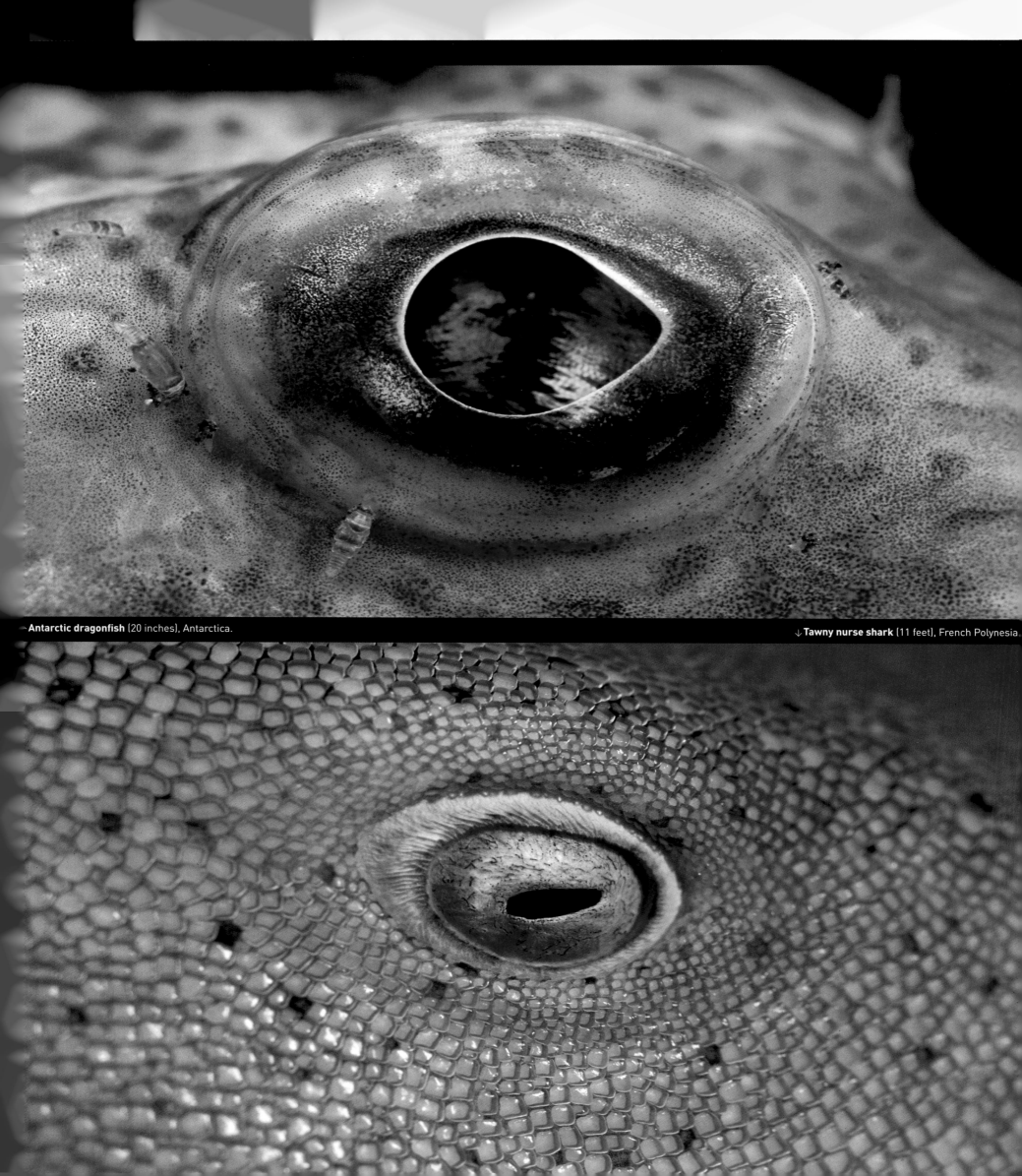

Antarctic dragonfish (20 inches), Antarctica.

↓ **Tawny nurse shark** (11 feet), French Polynesia.

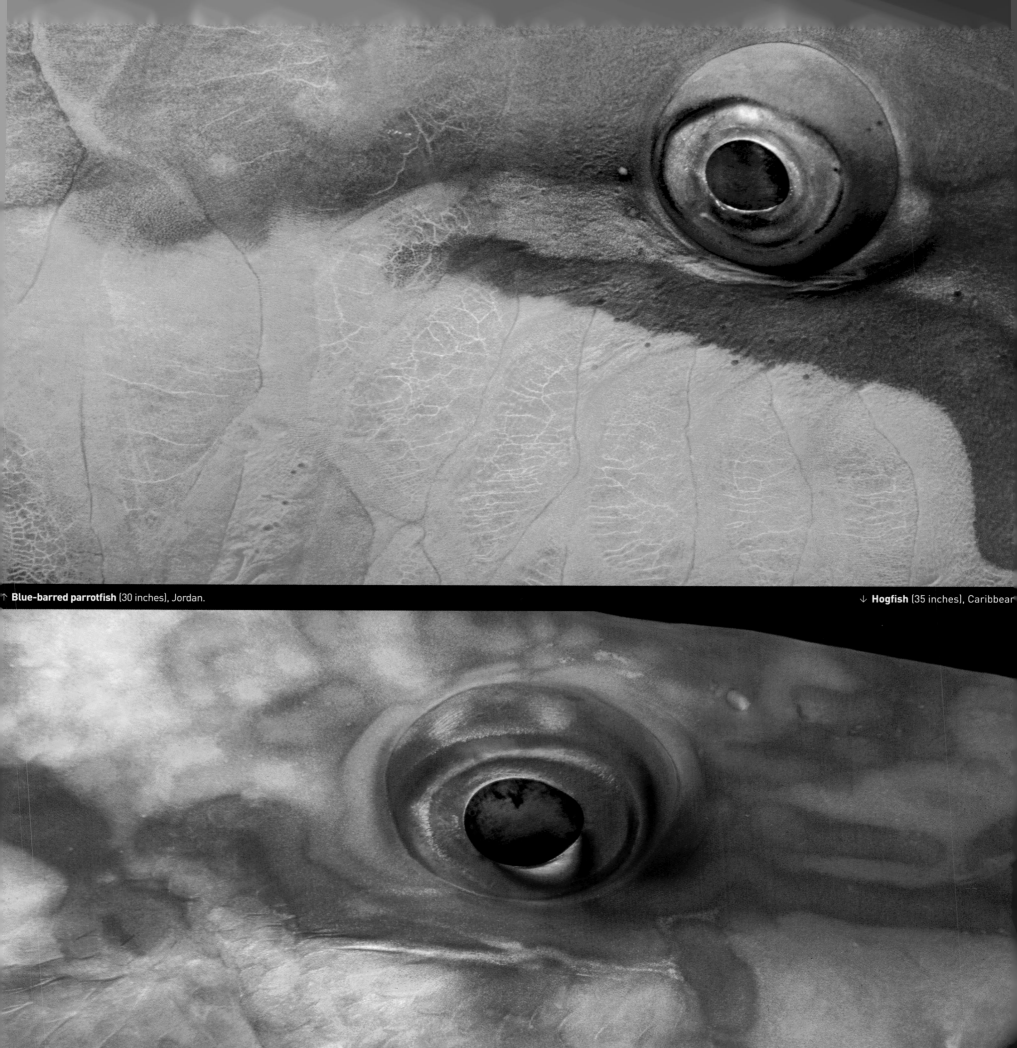

↑ **Blue-barred parrotfish** (30 inches), Jordan.

↓ **Hogfish** (35 inches), Caribbean

Staghorn coral
(6 feet), Papua New Guinea.

Space Where There's Freedom

Mediterranean morays (4 feet), France.

At sea, everyone can have the experience of freedom: infinite blue horizons for the sailor,
the sense of weightlessness for the skin diver, contact with wild fauna for the undersea diver.
The oceans are immense and harbor the dreams of humanity. This otherworld without fences
and billboards is nothing less than the last wild continent—and we need it.

The Deep Seas: An Unknown World

The deep sea is the most mysterious and poorly understood ecosystem on Earth. It includes all the oceans below the "light zone" and begins with the twilight world of the mesopelagic[1] realm.

Bioluminous wonders of the deep

Many animals that live in this world of twilight have special adaptations, including highly sensitive eyes and dark or mirrorlike bodies for camouflage. They also have luminous organs that help to break up the silhouette of the animal against downwelling light. These light organs may also be used for attracting prey, signaling to

Deeper still

Below the twilight zone, the abundance and diversity of species gradually diminishes as we enter the permanently dark bathypelagic[2] realm, where life is relatively sluggish. Here animals are generally black, red, dark brown, or purple to absorb bioluminescent light and hide from predators. Residents of these poorly known depths include voracious scavenging crustaceans and deep-sea anglerfish with enormous teeth (compared to their body size), and equipped with luminescent lures to trap other animals.

relatively featureless it harbors a huge diversity of smaller animals, including worms, tiny bivalve mollusks, and copepod crustaceans. Scientists have only sampled the equivalent of several football fields of deep-sea mud, and one of the great puzzles for deep-sea biologists is just how many species live in the mud of the abyssal plains. Estimates range from 1 million to 100 million or more species and we do not know how far across the ocean floor any one species is distributed. This means that it is not possible to predict how likely it is that future activities, such as deep-sea mining, will cause extinctions of deep-sea species.

A variety of seascapes

The abyssal plains are not the only seabed habitat in the deep sea. Canyons divide the continental slopes and funnel food from the continental shelves into the deep sea. Abundant communities of animals live on the seabed and fish and other predators swarm within or above the canyons. The oceans are also crisscrossed with submarine mountain ranges, plateaus, and drowned micro-continents. These features all interact with deep-ocean currents, injecting nutrient-rich deep-sea water into the light zone and sometimes causing an increase in surface productivity. They also trap the daily migrating layers of plankton from the twilight zone and focus food supply over a very small area of the ocean. Many of these features are hotspots of biological activity in the desertlike oceans of the high seas. As such, they attract high concentrations of spectacular predators such as whales and sharks and host populations of specially adapted deep-sea fish such as orange roughy, alfonsinos or armorheads.

Volcanic phenomena: a dark energy

Many of the mountain ranges in the deep oceans have been generated by volcanic processes, especially when new ocean crust is being generated, where plates that form the surface of Earth are moving apart. In these same areas, cracks that form in the newly created seabed allow seawater to seep down and come into contact with rocks heated by the lava beneath. The water is heated to high temperatures and becomes enriched with chemicals from the rocks, and because it becomes buoyant, shoots out of the seabed through hydrothermal vents. Some of these vents appear as chimneys of minerals spouting "black smoke," as the minerals in the hydrothermal waters precipitate in the cold water of the bottom of the ocean. This rich chemical soup is an energy source that allows unique communities of giant tube-worms, clams, mussels, and shrimp to thrive without depending on energy from the sun. Communities based on this "dark energy" are unique and have

The abyssal environment covers two-thirds of Earth's surface. Worldwide, numerous stocks of "deep" fish have already been overexploited. These species regenerate slowly and can only bear very limited exploitation rates.

members of the same species, or for producing defensive flashes of light to startle predators. Many animals in this zone undertake the world's largest daily migrations, up into the surface water layers at night to feed and then back into the twilight zone in the morning to avoid fast-swimming visual predators of the light zone.

Millions of species to discover

The deep sea also includes the bottom of the oceans, where the edges of the continental shelves gently slope onto the vast abyssal plains. Here, soft mud is pockmarked by the tracks and trails of large animals, such as sea cucumbers, giant sea spiders, and worms that sweep the surface clean of particles of food that rain down from the light zone. Despite the fact that deep-sea mud looks

changed the view of scientists on how life may have begun and on the possibility of life on other planets, where similar chemical energy sources may exist.

Special communities of deep-sea animals also live in other strange habitats that include parts of the continental margins where chemical-rich waters or methane gas leak out of the seabed, or around the carcasses of large, dead animals, such as whales. These special communities show evolutionary relationships to hydrothermal vents. Even in the deepest parts of the oceans—the trenches—life has been found and species that occur nowhere else live under huge pressures in total darkness up to six miles from the surface of the oceans.

Science needs to be a priority

Because of the rich food supply, many seamounts are home to rich and diverse communities of cold-water corals, gorgonians, and sponges. Many of the animals living in these habitats have not been sampled anywhere else in the world by scientists. Because seamounts host large populations of deep-sea fish, they have been targeted by fleets of trawlers operating in national waters and on the high seas. The fish that live on seamounts are extremely long-lived (more than 150 years in some cases), slow-growing, and slow to reproduce. They are therefore highly vulnerable to overfishing and many populations of these species have already crashed under the pressure from trawlers and long-liners. Moreover, technological advances in video imaging and photography of deep-sea ecosystems have also revealed the tragic consequences of trawling on the rich communities of deep-sea animals that live on seamounts and on other deep-sea habitats. The trawls smash many of the habitat-forming organisms to pieces, destroying deep-sea reefs and potentially extinguishing the unique species that live on seamounts. Habitats that have been destroyed by deep-sea trawling have not even been explored by scientists, so we do not know what has been lost already.

Alex Rogers was trained as a marine biologist at the University of Liverpool's Port Erin Marine Laboratory. He then undertook research fellowships at the Marine Biological Association, Plymouth and Southampton Oceanography Centre, where he developed an interest in deep-sea ecology, biodiversity, and molecular biology. Dr. Rogers is currently senior research fellow at the London Institute of Zoology. He works on deep-sea marine fauna and is an advisor to the IUCN Global Marine Program.

Sabertoothed viperfish.

THE DEEP ECOSYSTEMS OF THE MEDITERRANEAN

The Mediterranean Sea is a semienclosed sea, which, in spite of its small surface area, can reach depths of more than three miles. Its most important features for their influence on the ecology of its deep ecosystems are its high temperature and salinity, as well as the constant value of its temperature at around 55°F through the whole water column starting at a depth of 650 feet. Another feature of the Mediterranean, important for many oceanographic and biological processes, is the fact that it is formed by two basins separated by Sicily and communicating only at the surface by the Straits of Messina and the Sicily Canal. There are therefore few possibilities for genetic exchanges between the adjacent basins, especially for the deepwater species.

The high temperature causes a degradation of the organic material in the first portion of the water column, where the transfer of energy to the great depths is small. This is why the deeps of the Mediterranean Sea have little diversity and a relatively simple trophic structure. As in

the case of the oceans, the Mediterranean fauna become more and more impoverished with increased depth.

However, mouths of large rivers are on the narrow platform that constitutes the continental shelf, where there are deep undersea canyons that facilitate the transport of sediments from the coastal zone to the great depths. These structures are essential for understanding how coastal disturbances have a significant repercussion in the short and medium term on the deep ecosystems, for the flow of particles to the bottom as well as for the accumulation of physical/chemical waste and pollution in general.

Finally, the existence should be noted of practically unknown extreme ecosystems, such as the seamounts, brine lakes, or accumulations of gas, whose role in maintaining biodiversity is unknown.

Evidence shows that the deep ecosystems of the Mediterranean are very vulnerable to anthropogenic contributions (fishing and pollution,) as well as to climate change. It is therefore essential to implement rapid, effective measures for the conservation and durable development of all of these different ecosystems. In this context, the General Mediterranean Fishing Commission (FAO) adopted in 2005 the prohibition of fishing any deeper than 3,280 feet.

Francisco Sardá works at the Institute of Marine Sciences (CMIMA-CSIC), in the Department of Renewable Marine Resources, in Barcelona. He is a specialist in the deep ecosystems of the Mediterranean and their preservation.

[1] Water layer between 656 and 3,280 feet.

[2] Water layer between 3,280 and 13,123 feet; below starts the abyssopelagic layer.

Epilogue

Being a mariner is not limited to mastering an arsenal of techniques. A skipper's knowledge about luffing, the abilities of the fisher on the high seas, or the erudition of the diver who is proficient at Trimix mixtures are all of little importance. These skills are not inconsequential, but they don't count for much. Real mariners are those who love the sea for the sea's sake.

To be a marine biologist or an underwater photographer, or both, is to be an observer—starting to observe right from the pier, watching the mussels that colonize the hull of a ship, then, out at sea, observing the wahles that spout in the distance, far behind the wheelhouse.

Being a marine biologist also means to be a great dreamer. Our observations are only tiny windows opened into an unknown world where man is only an intruder.

Clumsy and shivering, we can only assess our unfitness for the aquatic life as soon as we are posessed with the desire to visit this universe. The return to the surface is always premature, and that's why doing scientific analysis—and dreaming—takes over when we are on land. The phrase so often heard, "Come on, please, come back down to earth!" has lead in its wings; it doesn't apply to us. Don't look for us on the moon. We're underwater.

**Descent along the slope of
the coral reef at Abu Qifan**
(1,650 feet tall), Egypt.

Exploration under the pack ice in the Canadian Arctic.

Dive to a depth of 460 feet on the sheer drop of a coral reef, Egypt.

Decompression stage after a dive using helium-based mixtures, Italy.

Whether we are skin diving, or diving with the most of modern diving suits, we are still subject to the ocean's rules.
Cold temperatures, water pressure, and decompression stops remind us that underwater, we are the intruders.
Of course, diving techniques and training methods are improving. Today, anyone can enjoy a dive for a longer and longer time, and can go deeper and deeper. And yet ... Our current state of evolution as underwater explorers is minuscule—nothing more than a thin strip of coastline compared to the immensity of the oceans. So, is it hopeless? Quite the contrary: We are only the third generation of scuba divers, and exploration has only just begun.

Diving

Being able to understand the oceans better provides added depth to underwater observation. From contemplator, we become investigator, and with this change in point of view, smaller and smaller things satisfy us. A naturalist splashing about at low tide is no doubt as happy as the novice seafarer in the waters at the end of the earth. Unfortunately, there are fewer and fewer biologists in the field. Biological research has become theoretical and molecular, while the field of ecology is becoming mathematical and based on modeling. This change may not always be for the good. Could Darwin have set forth his famous theory of evolution without visiting the Galapagos Islands?

Underwater tracking equipment used for monitoring shark movements French Polynesia.

Close-up of the equipment used for the underwater tracking of sharks, French Polynesia.

Studying the stocks of red coral, France.

Evaluating coral mortality, Jordan.

Collecting observations during a dive, Jordan.

Attaching an acoustic transmitter to the point of a harpoon to study sharks, French Polynesia.

Diving with the wolf-eel (6 feet), Canada.

Pierre in the wreck of the
Cedar Pride, cargo ship
260 feet long, Jordan.

Laurent in the submerged
mangrove forest,
Mozambique Channel.

Pierre and a Commerson's
frogfish (12 inches), Jordan.

Taking Photographs

Laurent and a young humpback whale (20 feet), French Polynesia.

Doing underwater photography requires, above all, resolving a paradox:
you need to take the time necessary to do it well, as for any naturalist activity, but
at the same time you have to accept the constraints of diving, where
time is counted down in minutes. But this task is not that burdensome,
because nothing is more exotic than an underwater photograph. And it also
undoubtedly serves a legitimate purpose: "out of sight, out of mind," so says the
proverb—the inhabitants of the undersea world well deserve to have pictures of
their fragility brought up to land. The sea needs interpreters!

The wreck of the *Heinkel*, German bomber, France.

A year ago we begged our editor to put off publishing this book for several years. "To keep your dreams, you must first lose your illusions" was our motto, and it seemed impossible to come to the end of this project so quickly. In the end, we had only nice surprises. We had forgotten that the oceans are so vast that they can contain our dreams a thousand times over. There's no doubt they can also contain yours...

Sea goldies (6 inches) and **pigmy sweepers** (3 inches), Jordan.

The Authors

Pierre and Laurent on assignment for the Aqaba Marine Park, Jordan, 2002.

Their first joint "expedition," Spain, 1994.

Scientific assignment on sharks, French Polynesia, 2004.

Electroencephalogram to determine medical suitability for professional diving.

By Mathieu Foulquié, marine biologist and professional diver, who has been a colleague of Pierre Descamp and Laurent Ballesta since the beginning

Planet Ocean is a great deal more than a book for Laurent Ballesta and Pierre Descamp. It is the result of five years of un-remitting work and about 15 years of informal exchanges on the sea and oceans. It would be impossible to talk about them without going into the inevitable common places inherent in the friendship that binds us together. Perhaps I can simply show the way in which I have followed their astonishing journey, from the time they met to the completion of this book. Like an episode of *The Persuaders!* about a pair of thirty-somethings with a Moleskine journal as thick as a phonebook, here is the story of two different dispositions, two competitors and two careers which, at the end of the day, seem to come together as one.

Long before they met, Laurent Ballesta and Pierre Descamp shared the same "sandbox," the wonderful playground of the Mediterranean. They both had grown up by the sea, and had dipped their little tootsies in it long before their swim fins. From crab fishing to patient observation of pink shrimp in the cracks of some rock, their passion was born there and never left them.

They only met long afterwards, at the Montpellier School of Science in 1993. University life represented for both of these aspiring Cousteaus a doorway to knowledge and the dream of becoming a marine biologist. These two always had a sense of reality. Isn't their motto: "To be able to keep your dreams, you must first lose your illusions"?

Their university career, however, was punctuated by many frustrations and disappointments because of the emphasis placed on the so-called modern sciences, such as genetics or molecular biology, to the detriment of the more "natural" sciences. This is far from the adventures of the *Calypso* (Jacques-Yves Cousteau's ship) or the *Beagle* (Darwin's ship), and French research in marine ecology was dying. Nevertheless, where many would become discouraged and abandon their childhood dreams, our "seagoing musketeers" found the energy to carry on patiently for years at the university in order to obtain a Master's in Biology of Populations and Ecosystems, followed by a Master's in Marine Ecology.

At the end of 1997, to fulfill their mandatory civil service duty to their country, Laurent and Pierre left France to serve overseas as Technical Assistance Volunteers. Pierre chose the Indian Ocean and the Mayotte Lagoon where, for 16 months, he worked in the Fish and Marine Environment Department. On the other side of the globe, Laurent joined the Marine Resources Department of French Polynesia on Rangiroa Atoll and devoted himself, between dives, to studying the larval and juvenile stages of reef fish.

At the same time, photography became a key factor in Laurent Ballesta's underwater forays. Upon returning from Polynesia, loaded with his files and slides, he met Nicolas Hulot in the middle of the Corsican scrub. Everyone—or almost everyone—knows what happened next: he left as a scientific adviser for his first Ushuaia Nature broadcast in New Zealand, the first of an incredible series of expeditions.

By the end of 1999, with a wealth of new experience, Pierre and Laurent founded the Œil d'Andromède at Montpellier University, a marvelous alternative to the boring basics of molecular biology and other productivity policies of the research departments.

Laurent Ballesta (left) and Pierre Descamp (right).

Laurent and the stonefish, Jordan, 2002.

Pierre and his father, on the day of his first dive, Palavas-les-Flots, 1982.

Laurent puffy from the cold after a 90-minute dive in 29°F water, Antarctic, 2005.

Pierre and his cumbersome photographic equipment. Assignment to develop the heritage of Polynesia, 2001.

Driven by a fierce desire to bring their passion to life, they dedicated themselves completely to the service of a science they resolutely wanted to see more "alive," more developed. Pictures were combined with scientific observation and field study completed the analytical work. Success was immediate, and work continued on from there. Undersea diving became their principal tool.

Laurent was one of the first Frenchmen to adopt the Inspiration, a closed circuit recycling suit with electronic mixture management which is a valuable asset for underwater photography and a revolutionary tool for scientific studies. Five years later, during a trip to the Red Sea in 2004, he would reach a depth of 460 feet with this equipment, thus opening a new avenue in the exploration of ocean floors. As for Pierre, he developed a true talent for ferreting out studies and securing financing for them. Their complementary skills were obvious and were their principal strength. Thanks to their mastery of photography as a tool and their skills in the field, they revisited the sensitive work of the biologist. "The only true science is knowledge of the facts," said Buffon. They seem to have made this adage their daily leitmotif by giving a new dimension to naturalist study, in direct line with their illustrious predecessors.

Always accompanied by his photographic equipment, Laurent traveled the world in quest of his own unknown territories: Papua New Guinea, Malaysia, Polynesia, Madagascar, Siberia, Canada, Baja California, the Galapagos Islands, Mongolia, Sudan, the Arctic, Australia, the Antarctic... Starting at the top in the audio-visual world, he is a three-time winner of the Antibes World Festival of Underwater Pictures historic Palme d'Or (2000, 2002 and 2004).

For his part, Pierre endeavored to represent the Œil d'Andromède at large international congresses on biodiversity and conservation of the marine environment. From these trips would be born a fruitful collaboration with The World Conservation Union (IUCN), involved in *Planet Ocean*. Pierre also carried out numerous environmental surveys in France and abroad, taking him from Guadeloupe to Saudi Arabia, also visiting the Mediterranean, Qatar, Jordan and Polynesia. While one of them was illustrating the most beautiful places on the planet, the other was working for their protection. A single imperative from that time on seemed to drive their unalloyed persistence: finding a field of expression suitable for their manifold talent. A first book, *De la source à la mer* [From the Spring to the Sea], was an homage to the little-known underwater wealth of the Languedoc-Roussillon. This was soon followed by a documentary, *Le Septième Ciel des requins gris* [Science of Shark Sex], that they wrote for Canal+ and France3 and which was produced by Cyril Tricot.

There is no doubt, however, that *Planet Ocean* is the finest result of their teamwork. It alone can summarize their philosophy of using the camera in the service of science. What better homage could be given to the synergy that, in the biological sense of the term, illustrates the cooperation of multiple skills in accomplishing a single function? For those who know them, it is clear that this book could not have been produced without the 12 years of incubation that preceded its publication.

Long life to you, my friends, and may the winds of adventure continue to fill the sails of your hopes.

Acknowledgments

We would like to thank all of the people, companies and institutions that contributed their help to us. We would like them to know that without them, none of this would have been possible.

We especially want to thank the following:

Mathieu Foulquié, a friend and regular collaborator, who coordinated the work of identifying the species.

Michel Lafon, who had faith in us from our very first meeting and patiently allowed our project to ripen for three years.

Aurèle Cariès, our editorial manager, whose vitality and drive put *Planet Ocean*, into orbit (!), immediately believing in the possibility of a national exhibition at the French Senate.

François Simard (of the IUCN Global Marine Program), who was in charge of creating the IUCN texts.

Nadège Duruflé, under the direction of Pascal Vandeputte, who did the artwork for *Planet Ocean*,.

Jean-Jacques Gasty, who patiently scanned and processed more than 500 slides, and Severine Urbin and Nicolas Dematté, who personally attended to the printing quality.

Our assistant Marie Guillot, and her two inseparable accomplices, for their vitality, their valuable daily help and their tenacity in exploring closets and filing cabinets full of slides.

Jean-Michel Cousteau, who contributed the foreword for the English edition.

The creation of *Planet Ocean* required many collaborations over six years. Thanks to all our colleagues and partners:

Equipment
Aqualung AQUA LUNG diving equipment; Bersub lighting; Topstar diving suits; AP-Valves Inspiration recycling suits; Nikon **Nikon** photographic and optical equipment; Seacam hyperbaric chambers; Extrem'Vision hyperbaric chamber; Fuji Velvia 50 and Kodak VS 100 film; the LTDP professional photo laboratory in Montpellier.

The Marine Protected Areas that welcomed us:
Cerbère-Banyuls Marine Nature Reserve (France), Port Cros National Park (France), Medes Islands Reserve (Spain), Aqaba Marine Peace Park (Jordan), Scandola Nature Reserve (France), S-Channel Reserve (Mayotte, France), Bouches de Bonifacio International Park (France).

Dive centers:
Carnon Diving School; Club Octopus, Palavas-les-Flots; Blue Dolphin Diving School, La Grande-Motte; La Palanquée Club, La Grande-Motte; Europlongée, Gruissan; Massilia Plongée, Marseille; Sensations Bleues, Marseille; El Rei del Mar, Estartit, Spain; Raie Manta Club, Rangiroa, Polynesia; Bora Bora Diving Center, Polynesia; Marquesas Diving Center, Nuku Hiva; Iti Diving International, Tahiti, Polynesia; Walindi Diving Resort, Papua New Guinea; Loloata Diving Center, Papua New Guinea. Royal Diving Center, Jordan; God's Pocket Diving Resort, Vancouver, Canada (many thanks to Bill and Annie). Deep Wreck expedition team (Aldo Ferucci). Special thanks to Étienne Bourgois and Romain Troublé for having welcomed us aboard the *Tara*, the famous polar schooner.

The Creation of Andromède Environnement:
A big thank-you to Jacques Bonnafé, president of the University of Montpellier 2 and to all those who allowed us to set up at the University: Alain Guilbot, François Bonhomme, Monique Vianey-Liaud, Jean-Pierre Quignard, Arnaud Martin and Jean-Paul Fernandez (European University Center).

Thanks to the pioneers: Dominique Sarda (DIREN LR); Jean-Paul Salasse (Les Écologistes de l'Euzière); Mathieu Geoffray; Jean-Daniel Collin; Cyril Frances.

Partners:
The Fondation Nicolas Hulot pour la Nature et l'Homme: Cécile Ostria; Jean Larivière; Annabelle Jaeger; Ministry of Fishing and Pearl Culture of Polynesia; GIE Tahiti Tourism; World Bank/GEF: Nicole Glineur; UNESCO: Elizabeth Wangari; Joint Union for the Development of Fishing and Protection of Marine Areas in the Gulf of Aigues-Mortes; Hérault Departmental House of the Environment; General Council of the Eastern Pyrenees; General Council of Hérault; Regional Council of Languedoc-Roussillon; Corsica Office of the Environment, Marianne Laudato; RMC Water Agency.

The World Conservation Union (IUCN): Julia Marton-Lefèvre, Carl Gustaf Lundin, Andrew Hurd, James Oliver, Imene Meliane, Ameer Abdulla, Gabriel Grimsditch and Sarah Gotheil, as well as all of the authors of the IUCN texts.

The students and collaborators of Andromède Environnement: Priscilla Dupont; Sophie Carteron, Nicolas Dalias, Aude Langevin, Xavier Pellier, Jérôme Bourjea, Germain Court. The neighbors of Tela Botanica.

Scientific partners: Paris Museum of Natural History: Alain Couté; College of Advanced Studies: René Galzin; IFREMER (French Research Institute for Exploitation of the Sea); University of Corsica: Gérard Pergent; Creocean: Olivier Lebrun, Éric Dutrieux and Sébastien Thorin; Marseille University of the Mediterranean U2: Bernard Thomassin; Jean-Georges Harmelin; CNRS photographic library: Marie Bacquet.

Scientists who participated in the identification of species: Patrick Louisy; Martin Thiel, Department of Ocean Sciences, Coquimbo, Chile. Dr. Catalina T. Pastor, National Patagonic Center, Biology and Management of Aquatic Resources Research Unit. Hwan Su Yoon, Department of Biological Sciences, University of Iowa. Bill Detrich, Northeastern University, Boston, MA. Philippe Béarez, National Museum of Natural History, "Ecology and Management of Biodiversity" Department. Joseph Poupin, French Naval Academy Research Institute. Marc Verlaque, CNRS Oceanology Center of Marseille; Philippe Martin; Xavier Boutolleau.

Audiovisual partners: Pascal Anciaux and all of the Ushuaia Nature team; Eau Sea Bleu Productions and in particular Cyril Tricot for the many projects we shared; "Le Lokal" and in particular Philippe Aussel, who produced the trailer for the book. Thanks also to Gil Kebaïli and Luc Marescot for their advice and friendship.

Those who taught us to dive, guided us, or simply helped us underwater: Jean-Pierre Montseny, Olivier Brissac, Jean-Daniel Colin, Alain Delmas, Olivier Boniface, Patrick and Fred Maxant, Yves Lefèvre and Eric Leborgne, François Brun and Patrick Tonolini, Laurent Piccione, Cyril Gressot, Cédric and Evelyne Verdier, Aldo Ferucci and Roberto Rinaldi, Eric Bahuet and Jean-Marc Belin, and especially Yann Hubert for the hundreds of dives we shared in 1998, the decisive year.

We would like to express our deepest appreciation to Mr. Christian Poncelet, President of the French Senate, and to his Chief of Staff, Alain Méar, who enthusiastically received our idea of a national exhibition on the theme of the oceans.

Special thanks from Laurent to Nicolas Hulot. "Thank you for allowing me to share in some of your adventures; thanks for the assurance of your support and for the subtlety of your advice. nothing is rarer or more exemplary to see united in one person the delicate marriage of conviction and humility."

Finally, thanks to our families who excused our absences and understood our doubts, especially Cécile and Albin Descamp and Carola Levasseur.

The World Conservation Union

Founded in 1948, the World Conservation Union (IUCN) is the world's largest and most important conservation network. The Union brings together more than 80 States and 110 government agencies, more than 800 non-governmental organizations (NGOs), and some 10,000 scientists and experts from 181 countries in a unique worldwide partnership. The Union's mission is to influence, encourage and assist societies throughout the world to conserve the integrity and diversity of nature and to ensure that any use of natural resources is equitable and ecologically sustainable.

IUCN is a multicultural, multilingual organization with 1000 staff located in 62 countries. Its headquarters are in Gland, Switzerland. The Union supports and develops cutting-edge environmental science; implements this research in field projects around the world; and then links both research and results to local, national, regional and global policy and laws by convening dialogues between governments, civil society and the private sector.

The goal of the IUCN Global Marine Programme is to improve significantly the conservation of marine biodiversity and the sustainable use of the natural resources from marine and coastal ecosystems throughout the world. To do this, its priorities are to improve knowledge concerning both ecosystems as well as socioeconomic matters, to promote mechanisms for financing conservation, and to develop better governance of the oceans and cooperation with everyone involved in working for the conservation of biodiversity and the sustainable use of marine resources.

The IUCN Global Marine Programme receives support from the Total Corporate Foundation for biodiversity and the sea.

Le Fondation Nicolas Hulot pour la Nature et l'Homme

Since its creation in 1990, the Fondation Nicolas Hulot pour la Nature et l'Homme [Nicolas Hulot Foundation for Nature and Mankind] has had the task of modifying individual and collective behavior in order to preserve our planet with respect to sustainable development.

An apolitical and non-sectarian NGO, the Foundation participates in the dissemination of knowledge about the ecological condition of our planet and implements all of the means at its disposal to convince as many as possible of the need to take action to slow down the impact of human activities.

The only recognized French public interest foundation totally dedicated to education about the environment, its goal is to train young people and adults to respect nature and become good ecocitizens. It makes the public aware of the wealth and fragility of our natural heritage and supports local initiatives on behalf of the environment.

Its actions are based on three major themes: water, eco-citizenship and biodiversity.

Recognized in educational and environmental circles, supported by well-known personalities and scientists within its Ecological Oversight Committee, the Foundation is also active internationally as a member of the International Union for the Conservation of Nature and Natural Resources (IUCN) and is a consulting NGO to the United Nation's Economic and Social Council.

Its initiatives are also being carried out in Belgium, Morocco and Senegal.

For more information: **www.fnh.org**

Laurent Ballesta & Pierre Descamp

Andromède Environnement
Université de Montpellier 2
163, rue Auguste-Broussonnet
34090 Montpellier
www.planete-mers.fr

Credits for Photographs

All photographs are by Laurent Ballesta, except for the following pictures:
Jean-Michel Bompar: pp. 332, 308;
Pierre Descamp: pp. 161, 206 (bottom), 250-251, 291 (bottom),
345 (bottom right), 353 (top right), 363; Marie Guillot and Cyril Frances: p. 381 (top);
Jean-Georges Harmelin: pp. 5, 162-163; Yann Hubert: pp. 290-291 (top);
Uwe Kils: p. 168 (bottom); Pascal Kobeh: p. 173; Eric Leborgne: p. 367;
Gil Kebaïli: pp. 89, 381 (right); Cyril Tricot: p. 366 (center).

CNRS Photothèque: Marie-Josèphe Chrétiennot-Dinet: pp. 164-165, 212 (bottom), 336-337.

AFP: Laurent Fiévet/STF: p. 243 (top),
Hotli Simanjuntak/PIG: p. 243 (bottom)
CORBIS: Mickael Prince: p. 41, Natalie Fobes: p. 61,
Dave G. Houser: p. 293, Ralph A. Clevenger: p. 72
HOA-QUI: Jean Hwasig: p. 60, Emmanuel Valentin: p. 134, Jacana/Francis Latreille: p. 88,
Jacana/Kurt Amsler: p. 333, Michel Renaudeau: p. 188, Age/Photostock/
Gonzalo Azumendi: p. 268, Explorer/J. Joffre: p. 339 (bottom), Photostock/José Fuste Raga: p. 276
PÊCHEUR D'IMAGES: Philip Plisson: pp. 107, 181
SIPA: EPA/Onome Ogueme: p. 160, Cavallera: p. 309
GETTY: National Geographic/Paul A. Zahl: p. 357,
Sambaphoto/Christiano Burmester: p. 356

Designed and Produced
VANDEPUB
by
Nadège Duruflé

Editorial Supervision
Aurèle Cariès

Staff for English Edition

Project Editor: Lisa Thomas
Art Director: Peggy Archambault
Photo Editor: Olivier Picard
Text Editor: Donnali Fifield
Consultant: Arlo Hemphill
Copy Editor: Erica Rose

Printed in France

ISBN
978-1-4262-0186-8